Lecture Notes
in Business Information Processing 253

Series Editors

Wil van der Aalst
Eindhoven Technical University, Eindhoven, The Netherlands
John Mylopoulos
University of Trento, Povo, Italy
Michael Rosemann
Queensland University of Technology, Brisbane, QLD, Australia
Michael J. Shaw
University of Illinois, Urbana-Champaign, IL, USA
Clemens Szyperski
Microsoft Research, Redmond, WA, USA

More information about this series at http://www.springer.com/series/7911

Esteban Zimányi · Alberto Abelló (Eds.)

Business Intelligence

5th European Summer School, eBISS 2015
Barcelona, Spain, July 5–10, 2015
Tutorial Lectures

 Springer

Editors

Esteban Zimányi
Department of Computer and Decision
    Engineering
Universite Libre de Bruxelles
Brussels
Belgium

Alberto Abelló
Database Technologies and Information
    Management Group
Universitat Politecnica de Cataluny
Barcelona
Spain

ISSN 1865-1348   ISSN 1865-1356 (electronic)
Lecture Notes in Business Information Processing
ISBN 978-3-319-39242-4   ISBN 978-3-319-39243-1 (eBook)
DOI 10.1007/978-3-319-39243-1

Library of Congress Control Number: 2016939568

Printed on acid-free paper

This Springer imprint is published by Springer Nature
The registered company is Springer International Publishing AG Switzerland

Preface

The 5th European Business Intelligence Summer School (eBISS 2015) took place in Barcelona, Spain, in July 2015. Tutorials were given by renowned experts and covered several recent topics in business intelligence. This volume contains the lecture notes of the summer school.

The first paper reports on the state of the art (including research and tools) in relational database schema evolution (with special emphasis on data warehouse evolution) and how this impacts the surrounding applications. It includes a discussion on query-rewriting techniques to adapt database client software to the changes.

The second paper compares RDF Data Cube Vocabulary (QB) with QB4OLAP, an extension to support OLAP analysis. The former is an initiative of W3C to publish multi-dimensional data on the Web in such a way that it can be linked to related data sets and concepts. The later extends it to explicit aggregation functions, hierarchies, and descriptive attributes. Such extension facilitates the definition of a user-friendly query language (QL) that allows OLAP users not familiar with SW concepts or languages to retrieve data without any knowledge of RDF or SPARQL (QL is automatically translated into SPARQL queries). Examples of usage of QB4OLAP on Eurostat data sets are provided.

Next, the third paper presents a concrete application domain for linked open data analysis, which is social information. It presents a proposal for modeling social (mainly textual) data, so that it can be analyzed with OLAP tools. The complexity of the problem justifies the proposal of a software architecture and methodology for the management of this kind of project. The feasibility of the proposed approach is analyzed in the context of two specific projects: one in the subject area of Italian politics, and another in the subject area of a large consumer goods company.

The fourth paper explores the possibility of including linked open data in analytical tasks. Traditional data warehousing has relied on internal data to enable decision making. Nevertheless, more recent big data trends have moved the focus to external data. Success stories based on the use of data coming from social networks are well known, but we can also benefit from publicly available semantically annotated data, which is growing fast mainly but not exclusively with governmental support. In any case, the integration of external data presents new challenges in terms of lack of structure, high heterogeneity, and poor quality.

The last paper discusses the feasibility and importance of deriving key performance indicator (KPI) calculations (i.e., aggregate queries) from their informal specifications. Since the majority of KPIs are process-oriented, process models (i.e., Petri nets) are used. Thus, seven different patterns are identified, which relate query elements to process models tasks.

In addition to the lectures corresponding to the papers described here, eBISS 2015 had two other lectures directly related to industry:

- Toni Cebrián from Enerbyte, Spain: "Time Series DBs and Streaming Algorithms"
- Wilinton Tenorio and Eduard Gil from ClearPeaks, Spain: "Life at ClearPeaks, An Overview of the Most Relevant Projects"

These lectures are not included in this volume.

In this edition, eBISS joined forces with the Erasmus Mundus IT4BI-DC consortium and hosted its doctoral colloquium aiming at community building and promoting a corporate spirit among PhD candidates, advisors, and researchers of different organizations. The corresponding session, organized in two parallel tracks, included eight presentations, as follows:

- Waqas Ahmed, Pakistan: "Modeling Data Warehouses with Multiversion and Temporal Functionality"
- Nurefsan Gur, Turkey: "Business Intelligence over Linked Open Spatio-Temporal Data"
- Dilshod Ibragimov, Uzbekistan: "OLAP over Distributed RDF Sources"
- Azadeh Nasiri, Iran: "Requirements Engineering for Big Data Predictive Analytics"
- Bijay Neupane, Nepal: "Intelligence Detection and Prediction of Energy at the Device Level"
- Kasun Parera, Sri Lanka: "Model-Based Database Systems"
- Vasileios Theodorou, Greece: "Automating User-Centered Design of Data-Intensive Processes"
- Jovan Varga, Serbia: "Discovering Analytical Concepts from User Profiles"

We would like to thank the attendees of the summer school for their active participation, as well as the speakers and their co-authors for the high quality of their contribution in a constantly evolving and highly competitive domain. Finally, the lectures in this volume benefited greatly from the comments of the external reviewers.

March 2016   Esteban Zimányi
                                                                            Alberto Abelló

Organization

The 5th European Business Intelligence Summer School (eBISS 2015) was organized by the Department of Computer and Decision Engineering (CoDE) of the Université Libre de Bruxelles, Belgium, and the Database Technologies and Information Management Group of the Universitat Politècnica de Catalunya.

Program Committee

Alberto Abelló	Universitat Politècnica de Catalunya, BarcelonaTech, Spain
Marie-Aude Aufaure	Ecole Centrale de Paris, France
Ralf-Detlef Kutsche	Technische Universität Berlin, Germany
Patrick Marcel	Université François Rabelais de Tours, France
Esteban Zimányi	Université Libre de Bruxelles, Belgique

External Reviewers

Ahmed Ahmedov	Technische Universität Dresden, Germany
Waqas Ahmed	Université Libre de Bruxelles, Belgium
Gastón Bakkalian	Poznan University of Technology, Poland
Besim Bilalli	Universitat Politècnica de Catalunya, Spain
Dilshod Ibragimov	Université Libre de Bruxelles, Belgium
Mohammed Idris	Université Libre de Bruxelles, Belgium
Faisal Orakzai	Université Libre de Bruxelles, Belgium
Mahmoud Sakr	Université Libre de Bruxelles, Belgium
Muhammad Aamir Saleem	Aalborg University, Denmark
Vasileios Theodorou	Universitat Politècnica de Catalunya
Martin Ugarte	Université Libre de Bruxelles, Belgium

Contents

Schema Evolution for Databases and Data Warehouses

Petros Manousis[1], Panos Vassiliadis[1(✉)], Apostolos Zarras[1],
and George Papastefanatos[2]

[1] Department of Computer Science,
University of Ioannina (Ioannina, Hellas), Ioannina, Greece
{pmanousi,pvassil,zarras}@cs.uoi.gr
[2] Athena Research Center (Athens, Hellas), Athens, Greece
gpapas@imis.athena-innovation.gr

Abstract. Like all software systems, databases are subject to evolution as time passes. The impact of this evolution is tremendous as every change to the schema of a database affects the syntactic correctness and the semantic validity of all the surrounding applications and de facto necessitates their maintenance in order to remove errors from their source code. This survey provides a walk-through on different approaches to the problem of handling database and data warehouse schema evolution. The areas covered include (a) published case studies with statistical information on database evolution, (b) techniques for managing schema and view evolution, (c) techniques pertaining to the area of data warehouses, and, (d) prospects for future research.

1 Introduction

Evolution of software and data is a fundamental aspect of their lifecycle. In the case of data management, evolution concerns changes in the contents of a database and, most importantly, in its schema. Database evolution can concern (a) changes in the operational environment of the database, (b) changes in the content of the databases as time passes by, and (c) changes in the internal structure, or *schema*, of the database. *Schema evolution*, itself, can be addressed at (a) the conceptual level, where the understanding of the problem domain and its representation via an ER schema evolves, (b) the logical level, where the main constructs of the database structure evolve (for example, relations and views in the relational area, classes in the object-oriented database area, or (XML) elements in the XML/semi-structured area), and, (c) the physical level, involving data placement and partitioning, indexing, compression, archiving etc.

In this survey, we will focus on the evolution of the logical schema of relational data and also extend our survey to the special case of data warehouse evolution. For the rest, we refer the interested reader to the following very interesting surveys. First, it is worth visiting a survey by Roddick [1], which appeared 20 years ago and summarizes the state of the art of the time in the areas of schema versioning and evolution, with emphasis to the modeling, architectural and query

© Springer International Publishing Switzerland 2016
E. Zimányi and A. Abelló (Eds.): eBISS 2015, LNBIP 253, pp. 1–31, 2016.
DOI: 10.1007/978-3-319-39243-1_1

language issues related to the support of evolving schemata in database systems. Second, 16 years later, a comprehensive survey by Hartung, Terwilliger and Rahm [2] appeared, in which the authors classify the related tools and research efforts in the following subareas: (a) the management of the evolution of relational database schemata, (b) the evolution of collections of XML documents, and (c) the evolution of ontologies. In the web site http://dbs.uni-leipzig.de/en/publications one may also find a comprehensive list of publications in the broader area of schema and data evolution. From our part, the material that we survey is collected by exploiting three sources of information: (a) our own monitoring of the field over the years, (b) by building on top of the aforementioned surveys, and, (c) by inspecting the main database and business intelligence venues in the last years, to identify the new works that have taken place since the last survey.

We organize the presentation of the material as follows. In Sect. 2, we discuss empirical studies in the area of database evolution. In Sect. 3, we present the sate of practice. In Sect. 4, we cover issues related to the identification of the impact that database evolution has to external applications and queries, as well as to views. In Sect. 5, we cover the specific area of data warehouses from the viewpoint of evolution. We conclude with thoughts around open issues in the research agenda in the area of evolution in Sect. 6.

2   Empirical Studies on Database Evolution

In this section, we survey empirical studies in the area of database evolution. These studies monitor the history of changes and report on statistical properties and recurring phenomena. In our coverage we will follow a chronological order, which also allows us to put the studies in the context of their time.

2.1   Statistical Profiling of Database Evolution via Real-World Studies

Studies During the 1990's. The first account of a sizable empirical study, by Sjoberg [3], discusses the evolution of the database schema of a health management system over a period of 18 months, monitored by a tool specifically constructed for this purpose. A single database schema was examined, and, interestingly, the monitored system was accompanied by a metadata dictionary that allowed to trace how the queries of the applications surrounding the database relate to the tables and attributes of the evolving database. Specific numbers for the evolution of the system, during this period of 18 months, include:

- There was a 139% increase of the number of tables and a 274% increase of the number of attributes (including affected attributes due to table evolution), too.
- All (100%) the tables were affected by the evolution process.
- Additions were more than deletions, by an 28% tables and a 42% for attributes.

- An insignificant percentage of alterations involved renaming of relations or merge/split of tables.
- Changes in the type of fields (i.e., data type, not null, unique constraints) proved to be equal to additions (31 both) and somehow less than deletions (48) for a period of 12 months, during which this kind of changes were studied.
- On average, each relation addition resulted in 19 changes in the code of the application software. At the same time, a relation deletion produced 59.5 changes in the application code. The respective numbers for attributes were 2 changes for attribute additions and 3.25 changes for attribute deletions, respectively.
- The evolution process was characterized by an inflating period (during construction) where almost all changes were additions, and a subsequent period where additions and deletions where balanced.

Revival in Late 2000's. In terms of empirical studies, and to the best of our knowledge, no developments took place for the next 15 years. This can be easily attributed to the fact that the research community would find it very hard to obtain access to monitor database schemata for an in-depth and long study. The proliferation of free and open-source software changed this situation. So, in the last few years, there are more empirical studies in the area that report on how schemata of databases related to open source software have evolved.

The first of these studies came fifteen years later after the study of Sjoberg. The authors of [4] made an analysis on the database back-end of MediaWiki, the software that powers Wikipedia. The study conducted over the versions of four years, and came with several important findings. The study reports an increase of 100 % in the number of tables and a 142 % in the number of attributes. Furthermore, 41.5 % of the attributes of the original database were removed from the database schema, and 25.1 % of the attributes were renamed respectively. The major reasons for these alterations were (a) the improvement of performance, which in many cases induces partitioning of existing tables, creation of materialized views, etc., (b) the addition of new features which induces the enrichment of the data model with new entities, and (c) the growing need for preservation of database content history. A very interesting observation is that around 45% of changes do not affect the information capacity of the schema, but they are rather index adjustments, documentation, etc. A statistical study of change breakdown revealed that attribute addition is the most common alteration, with 39 % of changes, attribute deletion follows with 26 %, attribute rename was up to the 16 % and table creation involved a 9 % of the entire set of recorded changes. The rest of the percentages were insignificant.

Special mention should be made to this line of research [5], as the people involved in this line of research should be credited for providing a large collection of links[1] for open source projects that include database support. Also, it is worth mentioning here that the effort is related to PRISM (later re-engineered to PRISM++ [6]), a change management tool, that provides a language of Schema

---

[1] http://yellowstone.cs.ucla.edu/schema-evolution/index.php/Benchmark_Extension.

Modification Operations (SMO) (that model the creation, renaming and deletion of tables and attributes, and their merging and partitioning) to express schema changes (see Sect. 4.1 for details).

Shortly after, two studies from the Univ. of Riverside appear. In [7], Lin and Neamtiu study two aspects of database evolution and their effect to surrounding applications. The first part of the study concerns the impact that schema evolution has on the surrounding applications. The authors work with two cases, specifically the evolution of Mozilla, between 2005 and 2009 and the evolution of the Monotone version control system between 2003 and 2009, both of which use a database to store necessary information for their correct operation. The authors document and exemplify how the developers of the two systems address the issue of schema evolution between different versions of their products. The authors also discuss the impact of erroneous database evolution, even though there exists software that is responsible for the migration of the system's modules to the new database schema. One very interesting finding is that although the applications can include a check on whether the database schema is synchronized to the appropriate version of the application code, this check is not omnipresent; thus, there exist cases where the application can operate on a different schema than the one of the underlying database, resulting in crashes or data loss. At the same time, the authors have measured the breakdown of changes during the period that they have studied. The second part of the study concerns DBMS evolution (attention: *DBMS*, not database) from the viewpoint of file storage. The authors study SQLite, MySQL and Postgres on how different releases come with different file formats and how usable old formats can be under a new release of the DBMS. Also, the authors discuss how the migration of stored databases should be performed whenever the DBMS is upgraded, due to the non-compatibility of the file formats of the different releases.

In a similar vein, in [8], Wu and Neamtiu considered 4 case studies of embedded databases (i.e., databases tightly coupled with corresponding applications that rely on them) and studied the different kinds of changes that occurred in these cases. Specifically, the authors study the evolution of Firefox between 2004 and 2008, Monotone (a version management system) between 2003 and 2010, BiblioteQ (a catalog management suite) between 2008 and 2010 and Vienna (an RSS newsreader) between 2005 and 2010. Comparing their results to previous works, the authors see the same percentages concerning the expansion of the database, but a larger number of table and column deletions. This is attributed to the nature of the databases, as the databases that are studied by Wu and Neamtiu are embedded within applications, rather than largely used databases as in the case of the previous studies. Moreover, the authors performed a respective frequency and timing analysis, which showed that the database schemata tend to stabilize over time, as the evolution activity calms down over time. There is more change activity for the schemata at the beginning of their history, whereas the schemata seem to converge to a relatively fixed structure at later versions.

A Large-Scale Study in 2013. In [9], Qiu, Li and Su report on their study of the evolution of 10 databases, supporting open source projects. The authors collected the source files of the applications via their SVN repositories and isolated the changes to the logical schema of each database (i.e., they ignored changes involving comments, syntax correction, DBMS-related changes, and several others). The remaining changes are characterized by the authors as valid DB revisions. The authors report that they have avoided the automatic extraction of changes, as the automatic extraction misses changes like table split or merge, or renaming and have performed manual checks for all the valid DB revisions for all the datasets. The study covers 24 types of change including the additions and deletions of tables, attributes, views, keys, foreign keys, triggers, indexes, stored procedures, default value and not null constraints, as well as the renaming of tables, attributes and the change of data types and default values. We summarize the main findings of the study in four categories.

Temporal and Locality Focus. Change is focused both (a) with respect to time and (b) with respect to the tables that change. Concerning timing, a very important finding is that 7 out of 10 databases reached 60% of their schema size within 20% of their early lifetime. *Change is frequent in the early stages of the databases, with inflationary characteristics; then, the schema evolution process calms down.* Schema changes are also focused with respect to the tables that change: 40% of tables do not undergo any change at all, and 60%-90% of changes pertain to 20% of the tables (in other words, 80% of the tables live quiet lives). The most frequently modified tables attract 80% of the changes.

Change breakdown. The breakdown of changes revealed the following catholic patterns: (a) insertions are more than updates which are more than deletions and (b) table additions, column additions and data type changes are the most frequent types of change.

Schema and Application Co-evolution. To assess how applications and databases co-evolve, the authors have randomly sampled 10% of the valid database revisions and manually analyzed co-evolution. The most important findings of the study are as follows:

- First, the authors characterized the co-change of applications in four categories and assessed the breakdown of changes per category. In 16.22% of occasions, the code change was in a previous/subsequent version than the one where the database schema change occurred; 50.67% of application adaptation changes took place in the same revision with the database change, 21.62% of database changes were not followed by code adaptation and 11.49% of code changes were unrelated to the database evolution.
- A second result says that each atomic change at the schema level is estimated to result in 10 − 100 lines of application code been updated. At the same time, a valid database revision results in 100 − 1000 lines of application code being updated.

A final note: Early in the analysis of results, the authors claim that change is frequent in schema evolution of the studied datasets. Although we do not dispute the numbers of the study, we disagree with this interpretation: change caries a lot between different cases (e.g., coppermine comes with 8.3 changes and 14.2 atomic changes per year contrasted to 65.5 changes and 299.3 atomic changes per year at Prestashop). We would argue that change can be arbitrary depending on the case; in fact, each database seems to present its own change profile.

2.2 Recent Advances in Uncovering Patterns in the Evolution of Databases

A recent line of research that includes [10–12], reveals patterns and regularities in the evolution of database schemata. At a glance, all these efforts analyze the evolution of the database schemata of 8 open source case studies. For each case study, the authors identified the changes that have been performed in subsequent schema versions and re-constructed the overall evolution history of the schema, based on Hecate, an automated change tracking tool developed by the authors for this purpose. The number of versions that have been considered for the different schemata ranged from 84 to 528, giving a quite rich data set for further analysis. Then, in [10] the authors perform a macroscopic study on the evolution of database schemata. Specifically, in this study the authors detect patterns and regularities that concern the way that the database schema grows over time, the complexity of the schema, the maintenance actions that take place and so on. To detect these patterns they resort to the properties that are described in Lehnman's laws of software evolution [13]. In [11], extend their baseline work in [10] with further results and findings revealed by the study, as long as detailed discussions concerning the relevance of the Lehman's laws in the case of databases, and the metrics that have been employed. On the other hand, in [12] the authors perform a microscopic study that delves into the details of the life of tables, including the tables' birth, death, and the updates that occur in between. This study reveals patterns, regularities and relations concerning the aforementioned aspects.

The Life of a Database Schema. In the early 70's, Lehman and his colleagues initiated their study on the evolution of software systems [14] and continued to refine and extend it for more than 40 years [13]. Lehman's laws introduce the properties that govern the evolution of *E-type systems*, i.e., software systems that solve a problem, or address an application in the real world [13]. For a detailed historical survey of the evolution of Lehman's laws the interested reader can refer to [15]. The essence of Lehman's laws is that *the evolution of an E-type system is a controlled process that follows the behavior of a feedback-based mechanism.* In particular, the evolution is driven by *positive feedback* that reflects the need to adapt to the changing environment, by *adding functionalities* to the evolving system. The growth of the system is constrained by *negative feedback* that reflects the need to perform *maintenance activities*, so as to prevent the deterioration of the system's quality.

In more detail, as discussed in [10,11] the laws can be organized in three groups that concern different aspects of the overall software evolution process.

The first group of laws discusses the existence of the feedback mechanism that constrains the uncontrolled evolution of software. The second group focuses on the properties of the growth part of the system, i.e., the part of the evolution mechanism that accounts for positive feedback. Finally, the third group of laws discusses the properties of perfective maintenance that constrains the uncontrolled growth, i.e., the part of the evolution mechanism that accounts for negative feedback. The major patterns and regularities revealed in [10, 11] from the investigation of each group of laws are summarized below:

– *Feedback mechanism for schema evolution*: Overall, the authors found that schema evolution demonstrates the behavior of a stable, feedback-regulated system, as the need for expanding its information capacity to address user needs is controlled via perfective maintenance that retains quality; this antagonism restrains unordered expansion and brings stability. Positive feedback is manifested as expansion of the number of relations and attributes over time. At the same time, there is negative feedback too, manifested as house-cleaning of the schema for redundant attributes or restructuring to enhance schema quality. In [10, 11] the authors further observed that the inverse square models [16] for the prediction of size expansion hold for all the schemata that have been studied.
– *Growth of schema size due to positive feedback*: The size of the schema expands over time, albeit with versions of perfective maintenance due to the negative feedback. The expansion is mainly characterized by three patterns/phases, (i) abrupt change (positive and negative), (ii) smooth growth, and, (iii) calmness (meaning large periods of no change, or very small changes). The schema's growth mainly occurs with spikes oscillating between zero and non-zero values. Also, the changes are typically small, following a Zipfian distribution of occurrences, with high frequencies in deltas that involved small values of change, close to zero.
– *Schema maintenance due to negative feedback*: As stated in [11] the overall view of the authors is that due to the criticality of the database layer in the overall information system, maintenance is done with care. This is mainly reflected by the decrease of the schema size as well as the decrease in the activity rate and growth with age. Moreover, the authors observed that age results in a reduction of the complexity to the database schema. The interpretation of this observation is that perfective maintenance seems to do a really good job and complexity drops with age. Also, they authors point out that in the case of schema evolution, activity is typically less frequent with age.

The Life of a Table - Microscopic Viewpoint. In [12], the authors investigated in detail the relations between table schema size, duration and updates. The main findings of this study are summarized below:

– From a general perspective, early stages of the database life are more "active" in terms of births, deaths and updates, whereas, later, growth is still there, but deletions and updates become more concentrated and focused.

- The life and death of tables is governed by the *Gamma* pattern, which says that large-schema tables typically survive. Moreover, short-sized tables (with less than 10 attributes) are characterized by short durations. The deletions of these "narrow" tables typically take place early in the lifetime of the project either due to deletion or due to renaming (which is equivalent from the point of view of the applications: they crash in both cases).
- Concerning the amount of updates, most tables live quiet lives with few updates. The main reason is the dependency magnet phenomenon, i.e., table updates induce large impact on the surrounding dependent software.
- The relation between table duration and amount of updates is governed by the *inverse Gamma* pattern, which states that updates are not proportional to longevity, but rather, few top-changer tables attract most of the updates.
  - Top-changer tables live long, frequently they are created in the first version of the database and they can have large number of updates (both in absolute terms and as a normalized measure over their duration).
  - Interestingly top-changer tables, they are not necessarily the larger ones, but typically medium sized.

3   State of Practice

In this section, we discuss how the commercial database management systems handle schema changes. The systems that we survey are: (a) Oracle, (b) DB2 of IBM, and, (c) SQL Server and Visual Studio of Microsoft. Another part of this research is dedicated to the open sourced or academic tools that are dealing with the schema changes. Some of those tools are: (a) Django, (b) South, and, (c) Hecate.

3.1   Commercial Tools

Oracle - Change Management Pack (CMP). Oracle Change Management Pack ([17]) is part of Oracle Enterprise Manager. CMP enables the management and deployment of schema changes from development to production environments, as well as the identification of unplanned schema changes that potentially cause application errors.

CMP features the following concepts:

- Change plans: A change plan is an object that serves as a container for change requests.
- Baselines: A baseline is a group of database object definitions captured by the Create Baseline application at a particular point in time.
- Comparisons: A comparison identifies the differences found by the Oracle Change Management Pack in two sets of database object definitions that you have specified in the Compare Database Objects application.

The *Create Baseline* application enables users in creating database schema descriptions in a CMP format or plain SQL DDL files. These descriptions are used to compare, or make changes to other schemata.

The *Compare Database Objects* application allows DBA users to compare different "database" versions. This way, in case of an application error produced by a non-tested schema change applied in the database, the DBA can produce all changes a-posterior and find the cause of the application failure.

The *Synchronization Wizard* of CMP supports the user in modifying an item *target* to match another item *source*. The *Synchronization Wizard* needs a comparison of the *target* and *source* items, so it works after the *Compare Database Objects* application. The *Synchronization Wizard* orders the "transformation" steps, in order to produce the *target* item. This is, for example, to make sure that the foreign keys will be applied after the primary keys. Besides that, the *Synchronization Wizard* can delete items. This happens, when there is no *source* item. Moreover, if there is no *target* item, the *Synchronization Wizard* initially creates and then synchronizes a *new target* item with the *source* one. Finally, using the *Synchronization Wizard*, the user may keep or undo the changes made to a *target* item.

Another module that works similar to the *Synchronization Wizard* is the *DB Propagate* application of CMP, which allows the user to select one or more object definitions and reproduce them in one or more *target* schemata.

Two other applications of CMP are: *DB Quick Change*, and, *DB Alter*. The *DB Quick Change* application helps the user in making *one* change to a *single* database item. The *DB Alter* application helps the user in making one or more changes to one, or more database items (in comparison to the *Synchronization Wizard*, here there is no need of any preceding comparison).

Finally, the *Plan Editor* of CMP lets the user perform a single change plan on one or more databases, that he may keep or undo. The *Plan Editor* can perform a wider variety of changes, compared to those that *Synchronization Wizard*, *DB Alter*, *DB Quick Change*, and *DB Propagate* can perform. The *Plan Editor* allows the creation of a change plan that serves as a container for change requests (directives, scoped directives, exemplars, and modified exemplars), generates scripts for those change requests and executes them on one or more databases.

IBM - DB2. IBM DB2 provides a mechanism that checks the type of the schema changes [18] that the users want to perform in system-period temporal tables. A system-period temporal table is a table that maintains historical versions of its rows. A system-period temporal table uses columns that capture the begin and end times when the data in a row is valid and preserve historical versions of each table row whenever updates or deletes occur. In this way, queries have access to both current data, i.e., data with a valid current time, as well data from the past. Finally, DB2 offers the DB2 Object Comparison Tool [19]. It is used for identifying structural differences between two or more DB2 catalogs, DDL, or version files (even between objects with different names). Moreover, it is able to generate a list of changes in order to transform the *target* comparator into a new schema, described by the *source* comparator. Finally, it is capable to undo changes that were performed and committed in a version file, so as to restore it to a given previous version.

Temporal tables prohibit changes that result in loss of data. These changes can be produced by commands like DROP COLUMN, ADD COLUMN, and, ALTER COLUMN. All the changes, applicable to temporal tables, can be propagated back to the history of the schema, with only two exceptions, the renaming of a table and the renaming of an index.

Microsoft - SQL Server and Visual Studio. Change management support for Microsoft SQL Server comes with the SQL Server Management Studio [20] (SSMS). SSMS allows the user to browse, select, and manage any of the database objects (e.g., create a new database, alter an existing database schema, etc.) as well as visually examine and analyze query plans and optimize the database performance. SSMS provides data import export capabilities, as well as data generation features, so that users can perform validation tests on queries. Regarding the evolution point of view, it is capable of comparing two different database instances and returning their structural differences. The tool may also provide information on DDL operations that occurred, through the reports of schema changes. An example of such a report from [21] is displayed in Table 1.

Table 1. SSMS report

database name	start time	login name	user name	application name	ddl operation	object	type desc
msdb	2015-08-27 14:08:40.460	sa	sa	SSMS - Query	CREATE	dbo.DDL_History	USER_TABLE
TestDB	2015-08-26 11:32:19.703	sa	sa	SSMS	ALTER	dbo.SampleData	USER_TABLE

Another set of tools that Microsoft offers for the validation of SQL code is the SQL Server Data Tools [22] (SSDT). SSDT follows a project-based approach for the database schema and SQL source code that is embedded in the applications. A developer can use SSDT to locally check and debug SQL code (by using breakpoints in his SQL code).

Another tool that comes from Microsoft as an extension to Visual Studio is MoDEF [23]. MoDEF uses the model-view-controller idea for the database schema manipulation. In MoDEF, the user defines classes that represent the columns of a table in a relational database. The classes are mapped to relational tables that are created in the database via select-project queries. In MoDEF, the changes of the client model are translated to incremental changes as an upgrade script for the database. Moreover, MoDEF annotates the upgrade script with comments, to inform the DBA for the changes that are going to happen in the database schema.

3.2   Open Source or Academic Tools

Django. Likewise to MoDEF, Django [24] also uses the model-view-controller idea for the database schema manipulation. Regarding evolution, Django uses an automatic way to identify which columns were added or deleted from the tables between two versions of code and migrate these changes to the database schema. Django identifies the changes in the attributes of a class and then produces the appropriate SQL code that performs the changes to the underlying database schema.

South. South [25] is a tool operating on top of Django, identifying the changes in the Django's models and providing automatic migrations to match the changes. South supports five database backends (PostgreSQL, MySQL, SQLite, Microsoft SQL Server, and, Oracle), while Django officially supports four (PostgreSQL, MySQL, SQLite, and, Oracle). South also supports another five backends(SAP SQL Anywhere, IBM DB2, Microsoft SQL Server, Firebird, and, ODBC) through unofficial third party database connectors.

In South, one can express dependencies of table versions so as to have the correct execution order of migration steps and void inconsistencies. For example, in a case where a foreign key references a column that is not yet a key, this kind of problem can be identified and avoided.

The Autodetector part of South can extend the migrations that Django offers. Specifically, South can automatically identify the following schema modifications: model creation and deletion (create/drop a table), field changes (type change of columns) and unique changes, while Django can only identify the addition or deletion of columns.

Hecate. Hecate [26] is a tool that parses the DDL files of a project and compares the database schemata between versions. Hecate also exports the transitions between two versions, describing the additions and deletions that occurred between the versions (renames are treated as deletions followed by additions). Hecate also provides measures such as size and growth of the schema versions.

Hecataeus. Hecataeus [27] is a what-if analysis tool that facilitates the visualization and impact analysis of data-intensive software ecosystems. As these ecosystems include software modules that encompass queries accessing an underlying database, the tool represents the database schema along with its dependent views and queries as a uniform directed graph. The tool visualizes the entire ecosystem in a single representation and allows zooming in and out its parts. Most related to the topic of this survey, the tool enables the user to create hypothetical evolution events and examine their impact over the overall graph. Hecataeus does not simply flood the event over the underlying graph; it also allows users as to define "veto" rules that block the further propagation of an evolutionary event (e.g., because a developer is adamant in keeping the exact structure of a table employed by one of her applications). Hecataeus also rewrites the graph, after the application of the event so that both the syntactical and the semantic correctness of the affected queries and views are retained.

4   Techniques for Managing Database and View Evolution

In this section, we discuss the impact of changes in a database schema to the applications that are related to that schema. Given a set of scripts, the methods proposed in this part of the literature identify how database and software modules are affected by changes that occur at the database level. Techniques for query rewriting are also discussed. Closely to this topic is the topic of view adaptation: how must the definition (and the extent, in case of materialization) of a view adapt whenever the schema of its underlying relations changes?

4.1   Impact Assessment of Database Evolution

The problem of impact assessment for evolving databases has two facets: (a) the identification of the parts of applications that are affected by a change, and, (b) the automation of the rewritting of the affected queries, once they have been identified. In this subsection, we organize the discussion of related efforts in a way that reflects both the chronological and the thematic dimension of how research has unfolded. A summary of the different methods is given at the end of the subsection.

Early Attempts Towards Facilitating Impact Assessment. Maule, Emmerich and Rosenblum [28] propose a technique for the identification of the impact of relational database schema changes upon object-oriented applications. In order to avoid a high computational cost, the proposed technique uses slicing, so as to reduce the size of the program that is needed to be analyzed. At a first step, the authors use a prototype slicing implementation that helps them identify the database *queries* of the program. Then, with a data-flow analysis algorithm, the authors estimate the possible runtime values for the parameters of the query. Finally, the authors use an impact assessment tool, Crocopat, coming with a reasoning language (RML) to describe the impacts of a potential change to the stored data of the previous step. Depending on the type of change, a different RML program is run, and this eventually isolates the *lines of code of the program* that are related to the queries affected by the change. The authors evaluated their approach on a C# CMS project of 127000 lines of code, and a primary database schema of up to 101 tables, with 615 columns and 568 stored procedures. The experiments showed that the method needed about 2 min for each execution, where they found that there were no false negatives. On the other hand, there were false positives in the results, meaning that the tool was able to find all the lines of code that were affected, leaving none out, but also falsely reported that some lines of code would be affected, whilst this was not really happening.

Architecture Graphs. Papastefanatos et al. [29–31] introduced the idea of dealing with *both* the database and the application code in uniform way. The results of this line of research are grouped in the areas of (a) modeling, (b) change impact analysis, and (c) metrics for data intensive ecosystems (data intensive ecosystems are conglomerations of data repositories and the applications that depend on them for their operations). This line of work has been facilitated by the Hecataeus tool (see [29, 32]).

Concerning the modeling and the impact analysis parts, in [29], the authors proposed the use of the *Architecture Graph* for the modeling of data intensive ecosystems. The *Architecture Graph* is a directed graph where the nodes represent the entities of the ecosystem (relations, attributes, conditions, queries, views, group by clauses, etc.), while the edges represent the relationships of these entities (schema relationships, operand relationships, map-select relationships, from relationships, where relationships, group by relationships, etc.). In the same paper, the authors proposed an algorithm for the propagation of the changes of one entity to other related entities, using a *status* indicator of whether the imminent change is accepted, blocked or if the user of the tool should be asked.

In [30], the authors proposed an extension for the SQL query language, that introduced policies for the changes in the database schema. The users could define in the declaration of their database schema whether a change should be accepted, blocked or if the user should be prompted. In this work, the policies were defined over: (a) the database schema universally, (b) the *high level modules* (relations, views and queries) of the database schema, and, (c) the remaining entities of the database, such as attributes, constraints and conditions.

Regarding the metrics part, a first attempt to the problem was made by Papastefanatos et al., on ways to predict the maintenance effort and the assessment of the design of ETL flows of data warehouses under the prism of evolution in [31]. In [33], the same authors used a real world evolution scenario, which used the evolution of the Greek public sector's data warehouse maintaining information for farming and agricultural statistics. The experimental analysis of the authors is based in a six-month monitoring of seven real-world ETL scenarios that process the data of the statistical surveys. The Architecture Graph of the system was used as a provider of graph metrics. The findings of the study indicate that schema size and module complexity are important factors for the vulnerability of an ETL flow to changes.

In a later work [34], Manousis et al., redefine the model of the *Architecture Graph*. The paper extends the previous model by requiring the *high level modules* of the graph to include *input* and *output* schemata, in order to obtain an isolation layer that leads to the simplification of the policy language. The method is based on the annotation of modules with *policies* that regulate the propagation of events in the Architecture Graph; thus, a module can either block a change or adapt to it, depending on its policy. The method for impact assessment includes three steps that: (a) assess the impact of a change, (b) identify policy conflicts from different modules on the same change event, and (c) rewrite the modules to adapt to the change. It is noteworthy that simply flooding the evolution event over the *Architecture Graph* in order to assess the impact and perform rewrittings, is simply not enough, as different nodes can react with controversial policies to the same event. Thus, the three stages are necessary, with the middle one determining conflicts and a "cloning" method, for affect paths on the graph, in order to service conflicting requirements, whenever possible.

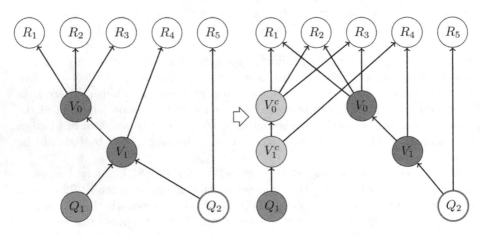

Fig. 1. A example of a rewrite process when the policies of Q_1 and Q_2 queries are conflicting [35].

In Fig. 1, we depict a situation that exemplifies the above. In the Architecture Graph that is displayed in the left part of Fig. 1, a change happens in view V_0 and affects the view V_1, which, in turn, affects the two queries Q_1 and Q_2 of the example. The first query (Q_1) accepts the change, whereas the second one (Q_2) blocks it. This means that Q_2 wants to retain its semantics and be defined over the old versions of the views of the Architecture Graph. Therefore, the query that accepted the change will get a new path, composed of "cloned", modified versions of the involved views that abide by the change (depicted in light color in the left part of the figure and annotated with a superscript c), whereas the original views and their path towards Q_2 retain their previous definition (i.e., they decline the change).

Schema Modification Operators. In this section, we review a work that produces –when it is possible– valid query rewritings of old queries over a new database schema, as if the evolution step of the database schema never happened. This way, the results that the user receives, after the execution of the rewritten query, are semantically correct.

An approach that supports the ecosystem idea, to a certain extent, is [36]. In this approach, the authors propose a method that rewrites queries whenever one of their underlying relations changes with the goal of retaining the same query result as if the evolution event never happened, using *Schema Modification Operators* (SMOs). The Schema Modification Operators that PRISM/PRISM++ tool uses are:

- CREATE TABLE $R(a, b, c)$
- DROP TABLE R
- RENAME TABLE R INTO T
- COPY TABLE R INTO T
- MERGE TABLE R, S INTO T
- PARTITION TABLE R INTO S WITH condition, T
- DECOMPOSE TABLE R INTO $S(a, b)$ $T(a, c)$
- JOIN TABLE R, S INTO T WHERE condition
- ADD COLUMN d [AS constant | function(a, b, c)] INTO R
- DROP COLUMN r FROM R
- RENAME COLUMN b IN R TO d

The R, S, and T variables represent relations. The a, b, c, d, and r variables represent attributes. The constant variable stands for a fixed value, while the function is used in ADD COLUMN in order to express simple tasks as data type and semantic conversions are. Besides the schema modification operators, PRISM/PRISM++ uses the integrity constraints modification operators ICMO and policies (which will be described later on) for this kind of rewrites. The ICMOs are:

- ALTER TABLE R ADD PRIMARY KEY $pk1(a, b)$ <policy>
- ALTER TABLE R ADD FOREIGN KEY $fk1(c, d)$ REFERENCES $T(a, b)$ <policy>
- ALTER TABLE R ADD VALUE CONSTRAINT $vc1(c, d)$ AS $R.e=$"0" <policy>
- ALTER TABLE R DROP PRIMARY KEY $pk1$
- ALTER TABLE R DROP FOREIGN KEY $fk1$
- ALTER TABLE R DROP VALUE CONSTRAINT $vc1$

The R and T variables represent relations. The a, b, c, d, and e variables represents attributes. The $pk1$ represents the primary key of the preceding relation. The $fk1$ represents the foreign key of a relation. Finally, the $vc1$ represents a value constraint. The ICMOs have, also, a <policy> placeholder, where the policy can be one of the following:

1. CHECK, where the PRISM/PRISM++ tool verifies that the current database satisfies the constraint, otherwise the ICMO is rolled back,
2. ENFORCE, where the tool removes all the data that violate the constraint, and,
3. IGNORE, where the tool ignores if there exist tuples that violate the constraint or not, but informs the user about this.

When the ENFORCE policy is used and tuples have to be removed, the tool creates a new database schema and inserts all the violating tuples in order to help the DBA carry out inconsistency resolution actions.

Regarding the rewrite process of queries through SMOs, the Chase &Backchase algorithm uses as input the SMOs and a query that is to be rewritten. The algorithm rewrites the query through an inversion step of the SMO's (for example, the inversion of a JOIN is the DECOMPOSITION), in order to retain the query's results unchanged, independently of the underlying schema. This way, the resulting tuples of the query will be the same as if the database schema never changed. The rewrite process of queries through ICMOs is done with the help of policies.

So, the steps that describe the algorithm of the rewriting that the authors proposed, are:

1. Get the SMOs from the DBA
2. Inverse the SMOs, in order to guarantee –if it is possible– the semantic correctness of the new query
3. Rewrite the query and validate its output.

The authors also describe a rewrite process of updates statement queries ("UPDATE table SET...") through SMOs and ICMOs, based in the ideas described in the previous paragraph. If the rewrite is through SMOs, the UpdateRewrite algorithm tries to invert the evolution step, while if the rewrite is through ICMOs, the policies ask the tool to check the tuples of the database and either guarantee or inform the user about the contents of the database.

To improve their rewrite time the authors try to minimize the input of the Chase &Backchase algorithm, by removing from the input all the mappings and constraints that are not related with the evolution step. Moreover, the proposed method uses only the version of the relation in which the query was written, leaving all the previous modifications out, as they are unrelated to the query. This is the backchase optimizer technique that the authors proposed, which produced bigger execution times in the chase and backchase phase of higher connected schemata because of the foreign keys that lead to higher input in chase phase, in the experiments that were conducted. In order to achieve even better execution time, the authors propose the use of a caching technique, since from the observations they made on their datasets, they noticed that there is a number of common query/update templates, which is parametrized and reused multiple times. These patterns are:

Join pattern type 1. In this pattern, a new table is created to host joined data from the desired column of two or more tables and migrates the data from the old tables to the new one.

Join pattern type 2. In this pattern, the data of a column are moved from the source table to the destination table.

Decompose pattern. In this pattern, a table is decomposed to two new tables. In order to be correct, both tables should have the key of the table.

Partition pattern. In this pattern, a part of the data of a table is moved into a new table and deleted from the original one.

Merge pattern. In this pattern, all the tuples of a table are moved into another table.

Copy pattern. In this pattern, an existing table is cloned.

The authors validated the PRISM/PRISM++ tool using the Ensembl project, including 412 schema versions, and the Mediawiki project, which is part of the Wikipedia project and had 323 schema versions. The authors used 120 SQL statements (queries and updates) from those two projects, tested them against SMO and ICMO operators and their tool found a correct rewriting, whenever one existed.

In a later work [6], the authors provide an extended description of the tool that performs the rewrites of the queries (PRISM/PRISM++) and its capabilities. Moreover, the authors introduce two other tools of which the first one collects and provides statistics on database schema changes and the other derives equivalent sequences of (SMOs) from the migration scripts that were used for the schema changes.

Summary. In Table 2 we summarize the problems and the solutions of the works that were presented earlier. The first two works are dealing with the impact analysis problem, which is to identify which parts of the code is affected by a change, and the other two works are dealing with the rewriting of the code in order to obtain or hide the schema changes.

4.2 Views: Rewriting Views in the Context of Evolution

A view is a query expression, stored in the database dictionary, which can be queried again, just as if it was a regular relation of the database. A view, thus, retains a dual nature: on the one hand, it is inherently a query expression; yet, on the other hand, it can also be treated as a relation. A *virtual view* operates as a macro: whenever used in a query expression, the query processor incorporates its definition in the query expression and the query is executed afterwards. *Materialized views* are a special category of views, that persistently store the results of the query in a persistent table of the DBMS. In this section, we survey research efforts that handle two problems. First, we start with the effect that a materialized view redefinition has on the maintenance of the view contents: the expression defining the view is altered and the stored contents of the view have to be adjusted to fit the new definition (ideally, without having to fully recompute the contents of the view from scratch). Second, we survey efforts pertaining to how views should be adapted when the schema of their defining tables evolves (also known as the "view adaptation" problem). A summary table concludes this subsection.

In [37], Mohania deals with the problem of maintaining the extent of a materialized view that is under redefinition, by proposing methods that try to avoid the full re-computation of the view. The author uses *expression trees*, which are binary trees, the leaf nodes represent base relations that are used for defining

Table 2. Summary table for Sect. 4.1

Works	Problem	Input	Output	Method
[28]	Impact analysis of an imminent schema change in OO apps	DB schema and source code; an imminent change	The lines of code that are affected by the DB schema change	Slicing technique to identify the DB related lines of C# code, and estimation of values so as to further slice the C# code.
[29,30]	Impact analysis of an imminent schema change	DB schema and application's queries abstracted as *Architecture Graph*; policies for of the nodes; an imminent change	Annotation of affected nodes with a status indication.	Language for node annotation. Propagation of a change, based on the node's *policy* for the change.
[34]	Restructuring of DB schema and app queries due to a schema change	DB schema and application's queries abstracted as *Architecture Graph*; policies for of the nodes; an imminent change	Rewritten *Architecture Graph* acording to the policies	Rewrite via cloning the queries that want to acquire the change and leave intact the ones that block the change.
[6,36]	Rewritting of app queries due to schema change	SMOs and ICMOs of the modification, and queries that use the modified table/view	Rewritten queries returning the same result as if the change has never happened.	The 1 hop away queries are rewritten as if the schema change never happened, using the *Chase &Backchase* algorithm

the view, while the rest of the nodes contain binary relational algebraic operators. Unary operators such as selection and projection are associated with the edges of the tree. In a nutshell, the author proposes that making use of these expression trees, it is easy to find common subexpressions between the new and old view statements and thus, if applicable, make use of the old view to get the desired results of the redefined view, without recomputing the new definition. Due to its structure, the tree allows to avoid interfering with the result of the view computation: (a) the height of the trees is no more than two levels, and, (b) a change is either a change to a unary operator associated with the edge of the tree, or a change to a binary node. This way, when the change is made at the root node, then the expression corresponding to the right hand child in the tree has to be evaluated only, while when the change is made at level $d=1$, the view re-computation becomes a view maintenance problem. Finally, when the change is made at any other node, it is only the intermediate results of the nodes that have to be maintained.

Gupta, Mumick, Rao and Ross [38] provide a technique that redefines a materialized view and adapts its extent, as a sequence of primitive local changes in the view definition, in order to avoid a full re-computation. Moreover, on more complex adaptations –when multiple simultaneous changes occur on a view– the local changes are pipelined in order to avoid intermediate creations of results of the materialized view. The following changes are supported as primitive local changes to view definitions:

1. Addition or deletion of an attribute in the SELECT clause.
2. Addition, deletion, or modification of a predicate in the WHERE clause (with and without aggregation).
3. Addition or deletion of a join operand (in the FROM clause), with associated equijoin predicates and attributes in the SELECT clause.
4. Addition or deletion of an attribute from the GROUP BY list.
5. Addition or deletion of an aggregate function to a GROUP BY view.
6. Addition, deletion or modification of a predicate in the HAVING clause. Addition of the first predicate of deletion of the last predicate corresponds to addition and deletion of the HAVING clause itself.
7. Addition of deletion of an operand to the UNION and EXCEPT operators.
8. Addition or deletion of the DISTINCT operator.

Concerning the problem of adapting a view definition to changes in the relations that define it, Nica, Lee and Rundensteiner [39] propose a method that makes legal rewritings of views affected by changes. The authors primarily deal with the case of relation deletion which (under their point of view) is the most difficult change of a database schema, since the addition of a relation, the addition of an attribute, the rename of a relation and the rename of an attribute can be handled in a straightforward way (the attribute deletion, according to the authors, is a simplified version of the relation deletion). To attain this goal one should find valid replacements for the affected components of the existing view, so, in order to achieve that, the authors of [39] keep a Meta-Knowledge Base on the join constraints of the database schema. This Meta-Knowledge Base (MKB) is modeled as a hyper-graph that keeps meta-information about attributes and their join equivalence attributes on other tables. The proposed algorithm, has as input the following: (a) a change in a relation, (b) MKB entities, and, (c) new MKB entities. Assuming that valid replacements exist, the system can automatically rewrite the view via a number of joins and provide the same output as if there was no deletion. The main steps of the algorithm are: (a) find all entities that are affected for Old MKB to became New MKB, (b) mark these entities and for each one of them find a replacement from Old MKB, using join equivalences, and, (c) rewrite the view over these replacements. Interestingly, the authors accompany their method with a language called E-SQL that annotates parts of a view (exported attributes, underlying relations and filters) with respect to two characteristics: (a) their dispensability (i.e., if the part can be

Table 3. Summary table for Sect. 4.2

Work	Problem	Input	Output	Method
[37]	Maintenance of redefined materialized views	Definition and redefinition of a materialized view	Recomputed content of the redefined view	Use of expression trees that identify common subexpressions between the input and output of their method, thus helping to avoid the full re-computation of a materialized view
[38]	Maintenance of redefined materialized views	Definition and redefinition of a materialized view	Recomputed content of the redefined view	The redefinition takes place as a sequence of primitive local changes (in complex adaptations this sequence is pipelined to avoid temporal results).
[39]	View adaptation on column deletion	Hypergraph that contains the join constraints of the DB schema	Valid replacement of column that is to be deleted	Search in the hypergraph (named MKB) for a replacement of the column that is to be deleted, and replace that column in the view with the replacement

removed from the view definition completely) and (b) their replaceability with an another equivalent part.

Summary. In Table 3 we summarize the problems and the solutions of the works that were presented earlier. The first two works refer to the problem of the recomputation of the contents of a materialized view, after a redefinition of the view. The other work refers to the problem of view adaptation on a column deletion in the source tables, via a replacement.

5   Techniques for Managing Data Warehouse Evolution

A research area where the problem of evolution has been investigated for many years is the area of data warehouses. In this section, we concentrate on works related on evolution of both schema and data modifications in the context of data warehouses, and we review methods and tools that help on the adaptation of those changes. We also refer the reader to two excellent surveys on the issue, specifically, [40, 41].

5.1   Data Warehouses and Views

At the beginning of data warehousing, people tended to believe that data warehouses were collections of materialized views, defined over sources. In this case, evolution is mostly an issue of adapting the views definitions whenever sources change.

Bellahsene, in two articles, [42, 43], proposed a language extension to annotate views with a HIDE clause that works oppositely to SELECT (i.e., the idea is to project all attributes except for the hidden ones and an ADD ATTRIBUTE clause to equip views with attributes not present in the sources (e.g., timestamps or calculations). Then, in the presence of an event that changes the schema of a data warehouse source (specifically, the events covered are attribute/relation addition and deletion), the methods proposed by the author for the adaptation of the warehouse handle the view rematerialization problems i.e., how to recompute the materialized extent via SQL commands. The author also proposes a cost model to estimate the cost of alternative options.

In [44], the author proposes an approach on data warehouse evolution based on a meta-model, that provides complementary metadata that track the history of changes (in detail, changes that are related to data warehouse views) and provide a set of consistency rules to enforce when a quality factor (actual measurement of a quality value) has to be re-evaluated.

5.2 Evolution of Multidimensional Models

Multidimensional models are tailored to treat the data warehouse as a collection of *cubes* and *dimensions*. Cubes represent clean, undisputed facts that are to be loaded from the sources, cleaned and transformed, and eventually queried by the client application, Cubes are defined over unambiguous, consolidated dimensions that uniquely and commonly define the context of the facts. Dimensions comprise *levels*, which form a hierarchy of degrees of detail according to which we can perform the grouping of facts. For example, the Time dimension can include the levels (1) Day, that can be rolled up to either (2a) Week or (2b) Month, both of which can be rolled up to level (3) Year. Each level comes with a domain of values that belong to it. The values of different levels are interrelated via rollup functions (e.g., 1/1/2015 can be rolled up to value 1/2015 at the Month level). As levels construct a hierarchy that typically takes the form of a lattice, evolution is mainly concerned with changing (i) the nodes of the lattice, or (ii) their relationship, or (iii) the values of the levels and their interrelationship. The problem that arises, then, is: how do we adapt our cubes (in their multidimensional form and possibly their relational representation) when the structure of their dimensions changes? The works surveyed in this subsection address this problem. A table at the end of the subsection summarizes the problems addressed and the solutions that are given.

The authors of [45] present a formal framework, based on a formal conceptual description of an evolution algebra, to describe evolutions of multi-dimensional schemata and their effects on the schema and on the instances. In [45], the authors propose a methodology that supports an automatic adaptation of the multi-dimensional schema and instances, independently of a given implementation. The main objectives of the proposed framework are: (i) the automatic adaptation of instances, (ii) the support for atomic and complex operations, (iii) the definition of semantics of evolution operations, (iv) the notification mechanism for upcoming changes, (v) the concurrent operation and atomicity of evolution operations, (vi) the set of strategies for the scheduling of effects and (vii) the support of the design and maintenance cycle.

The authors provide a minimal set of atomic evolution operations, which they use in order to present more complex operations. These operations are: (i) insert level, (ii) delete level, (iii) insert attribute, (iv) delete attribute, (v) connect attribute to dimension level, (vi) disconnect attribute from dimension level, (vii) connect attribute to fact, (viii) disconnect attribute from fact, (ix) insert classification relationship, (x) delete classification relationship, (xi) insert fact, (xii) delete fact, (xiii) insert dimension into fact, and, finally, (xiv) delete dimension.

In [46], the authors suggest a set of primitive dimension update operators that address the problems of: (i) adding a dimension level, above (generalize) or below (specialize) an existing level, (ii) deleting a level, (iii) adding or deleting a value from a level (add/delete instance), or (iv) adding (relate) or removing edges between parallel levels (unrelate). In [46], the authors also suggest another set of complex operators, that intend to capture common sequences of changes in instances of a dimension and encapsulate them in a single operation. The set of those operators consists of: (i) reclassify (used, for example, when new regions are assigned to salespersons as a result of marketing decisions of a company), (ii) split (used, for example, when a region is divided into more regions and more salespersons must be assigned to those regions due to the population density), (iii) merge (the opposite of split), and, (iv) update (used, for example, when a brand name for a set of items changes but the corporation as well as the set of products related to the brand remain unchanged).

The mappings that the authors propose, for the transitions from the multidimensional to the relational model, support both the de-normalized and normalized relational representations. In the de-normalized approach, the idea is to build a single table containing all the roll-ups in the dimension while in the normalized approach, the idea is to build a table for each direct roll-up in the dimension.

Finally, in the experiments that the authors conducted, they found that the structural update operators in the de-normalized representation are more expensive. The instance update operators in the normalized representation are more expensive because of the joins that have to be performed, whilst both representations are equally good for the operators that compute the net effect of updates.

In a later work, the authors of [47] suggest a set of operators which encapsulate common sequences of primitive dimension updates and define two mappings from the multidimensional to the relational model, suggesting a solution on the problem of multidimensional database adaptation.

The effects of evolution to alternative relational logical designs is explored in [48]. Specifically, the authors explore the impact of changes to both star and snowflake schemata. The changes covered include (i) the addition of deletion of attributes to levels, (ii) the addition/deletion of dimension levels, (iii) the addition/deletion of measures, and (iv) the addition/deletion of dimensions into fact tables. A notable, albeit expected, result is that comparison of the effect of changes to the two alternative structures, reveals that the simplest one, star schema, is more immune to change than the more complicated one.

Summary. In Table 4 we summarize the problems and the solutions of the research efforts that were presented earlier. The first two lines of work refer to the evolution of multidimensional database schemata and the adaptation of its contents, and the final effort refers to a comparison of the logical design between star and snowflake alternatives.

5.3   Multiversion Querying over Data Warehouses

Once the research community had obtained a basic understanding of how multidimensional schemata can be restructured, the next question that followed was: "what if we keep track of the history of all the versions of a data warehouse schema as it evolves?" Then, we can ask queries that span several versions

Table 4. Summary table for Sect. 5.2

Works	Problem	Input	Output	Method
[45]	Multidimensional database adaptation	MD schema; Changes of schemata	New schema and instances	Automatic adaptation of multi-dimensional schema and instances through simple and complex operators of an evolution algebra
[46, 47]	Multidimensional database adaptation	MD schema; Changes of schemata	Normalized or de-normalized new (RDBMS) schema	Use of primitive dimension update operators and complex operators that map the multidimensional schemata to RDBMS schemata
[48]	Evolution of alternative relational logical designs	Changes of schemata	Comparison of logical designs to changes	Perform the changes to both star and snowflake designs

having different structure, also known as *multiversion queries*. The essence of multi-version queries involves transforming the data of previous versions (that obey a previous structure) to the current version of the structure of the data warehouse, in order to allow their uniform querying with the current data.

In this section, we discuss the adaptation of multiversion data warehouses [49], the use of data mining techniques in order to detect structural changes in data warehouses [50–52], and, the use of graph representations (directed graphs) [53], in order to achieve correct cross version queries. We summarize problems and solutions in a table at the end of the subsection.

Eder and Koncilia [52] propose a multidimensional data model that allows the registration of temporal versions of dimension data in data warehouses. Mappings are provided to translate data between different temporal versions of instances of dimensions. This way, the system can answer correctly queries that span in periods where dimension data have changed. The paper makes no assumption on dimension levels, so when referring to a dimension, the paper implies a flat structure with a single domain. The mappings are described as transformation matrices. Each matrix is a mapping of data from version V_i to version V_{i+1} for a dimension D. Assume, for example a 2-dimensional cube, including dimensions A and B with domains $\{a_1, a_2\}$ and $\{b_1, b_2\}$ respectively. Assume that in a subsequent version: (i) a_1 is split to a_1^1 and a_1^2 and (ii) b_1 and b_2 are merged into a single value b. Then, there is a transformation matrix for dimension A, with one row for each old value $\{a_1, a_2\}$ and one column for each new value $\{a_1^1, a_1^2, a_2\}$ expressing how the previous values relate to the new ones. For example, one might say that a_1^1 takes 30 % of a_1 and a_1^2 takes the other 70 %. The respective matrix is there for dimension B. Then, by multiplying any cube with A and B as dimensions with the respective transformation matrices, we can transform an old cube defined over $\{a_1, a_2\} \times \{b_1, b_2\}$ to a new cube defined over $\{a_1^1, a_1^2, a_2\} \times \{b\}$.

So at the end, the resulting factual cube maps the data of the previous version to the dimension values of the current version; this way, both the current and the previous version can be presented uniformly to the user.

Eder, Koncilia and Mitsche [50] propose the use of data mining techniques for the detection of structural changes in data warehouses, in order to achieve correct results in multi-period data analysis OLAP queries. Making use of three basic operations (INSERT, UPDATE and DELETE), the authors are able to represent more complex operations such as: SPLIT, MERGE, CHANGE, MOVE, NEW-MEMBER, and DELETE-MEMBER. The authors propose several data mining techniques that detect which is the schema attribute that changed. In the experiments that were conducted, the authors observed that the quality of the results of the different methods depends on the quality and the volatility of the original data.

The same authors continue their previous work on data mining techniques for detection of changes in OLAP queries in [51]. Since their previous approach was incapable of detecting some variety of changes, the authors propose data mining techniques in form of multidimensional outlier detection to discover unexpected deviations in the fact data, which suggests that changes occurred in dimension data. By fixing a dimension member they get a simple two-dimensional matrix where the one axis is the excluded dimension member. From that matrix, a simple deviation matrix with relative differences is computed. In this deviation matrix, the results are normalized to get the probability of a structural change that might have occurred. The authors propose the 10 % as a probability threshold for the change to have occurred. From the conducted experiments, the authors found that this method analyzes the data in more detail and gives a better quality of the detected structural changes.

Some years later, Golfarelli et al. [53] propose a representation of data warehouse schemata as graphs. The proposed graph represents a data warehouse schema, in which the nodes are: (i) the fact tables of the data warehouse, and (ii) the attributes of fact tables (including properties and measures), while the edges represent simple functional dependencies defined over the nodes of the schema. The authors also define an algebra of schema graph modifications that are used to create new schema versions and discuss of how cross-version queries can be answered with the help of augmented data warehouse schemata. The authors finally show how a history of versions for data warehouse schemata is managed.

Since the authors' approach is based on a graph, the schema modification algebra uses four simple schema modification operations (M): (i) $\mathtt{Add}_F$ that adds an arc involving existing attributes, (ii) $\mathtt{Del}_F$ that deletes an existing arc, (iii) $\mathtt{Add}_A$ that adds a new attribute –directly connected by an arc to its fact node– and (iv) $\mathtt{Del}_A$ that deletes an existing attribute. Besides those simple operators, the authors define the $\mathtt{New}(S, M)$ operator that describes the creation of a new schema, based on the existing schema S when a simple schema modification M is applied.

The authors introduce augmented schemata to serve multiversion queries. Each previous version of the data warehouse schema is accompanied by an augmented schema whose purpose is to translate the old data under the old schema to the current version of the schema. To this end, the augmented schema keeps track of every new attribute (say A), or new functional dependency (say f). In order to translate the old data to the new version of the schema, the system might have to: (i) estimate values for A, (ii) disaggregate or aggregate measure values depending on the change of granularity, (iii) compute values for A, (iv) add values for A, or, (v) check if f holds.

The set of versions of the schemata is described by a triple (S, S^{AUG}, t), where S is a version, S^{AUG} is the related augmented schema and t is the start of the validity interval of S. This way, the history of the versions of the data warehouse can be described as a sequence of changes over changes, starting from the initial schema of the history: $H = S_0, S_0^{AUG}, t_0$. Since every previous version is accompanied by an augmented schema that transforms it to the current one, it is possible to pose a query that spans different versions and translate the data of the previous versions to a representation obeying the current schema, as explained above.

Practically around the same time, Wrembel and Bebel [49] deal both with cross-version querying and with the problems that appear when changes take place at the external data sources (EDS) of a data warehouse. Those problems can be related to a multi-version data warehouse which is composed of a sequence of persistent versions that describe the schema and data for a given period of time. The authors approach has a meta-data model with structures that support: (i) the monitoring of the external data sources on content and structural changes, (ii) the automated generation of processes monitoring external data sources, (iii) the adaptation of a data warehouse version to a set of discovered external changes, (iv) the description of the structure of every data warehouse version and (v) the querying of multiple data warehouse versions (cross version querying), and (vi) the presentation of the output as an integrated result.

The schema change operations that the authors support are: (i) the addition of a new attribute to a dimension level table, (ii) the removal of an attribute from a dimension level table, (iii) the creation of a new fact table, (iv) the association of a fact table with a dimension table, (v) the renaming of a table, and three more operations that are applicable to snowflake schemata, (vi) the creation of a new dimension level table with a given structure, (vii) the inclusion of a parent dimension level table into its child dimension level table, and, (viii) the creation of a parent dimension level table based on its child level table.

The instance change operations that the authors have worked on, are: (i) the insertion of a new level instance into a given level, (ii) the deletion of a level instance, (iii) the change of the association of a child level instance to another parent level instance, (iv) the merge of several instances of a given level into one instance of the same level, and (v) the split of a given level instance into multiple instances of the same level.

In order to query multiple versions, the authors' method is based on a simple and elegant idea: the original query is split to a set of single version queries. Then, for each single version query, the system does a best-effort approach: if, for example, attributes are missing from the previous version, the system omits them from the single version query; the system exploits the available metadata for renames; it can even, ignore a version, if the query is a group by query and the grouping is impossible. If possible, the collected results are integrated under the intersection of attributes common to all versions (if this is the case of the query); otherwise, they are presented as a set of results, each with its own metadata.

Regarding the detection of changes in external data sources, the authors propose a method that uses wrappers (software modules responsible for data model transformations). Each wrapper is connected to a monitor (software that detects predefined events at external data sources). When an event is detected, a set of actions is generated and stored in *data warehouse update register* in order to be applied to the next data warehouse version when the data warehouse administrator calls the warehouse refresher. The events are divided into two categories: (i) structure events (which describe a modification in the schema of the data warehouse) and (ii) data events (which describe a modification in the contents of a data warehouse). For each event, an administrator defines a set of actions to be performed in a particular data warehouse version. The actions are divided in two categories: (i) messages (which represent actions that cannot be automatically applied to a data warehouse version) and (ii) operations (for events whose outcomes can be automatically applied to a data warehouse version). Both categories of actions do not create a new data warehouse version automatically but require either the administrator to apply them *all* in an action definition of an explicitly selected version, or the actions are logged in a special structure for manual application of the ones the administrator wants to apply.

Summary. In Table 5 we summarize the problems and the solutions of the research efforts that were presented earlier. The first two lines of work refer to data translation between the versions of the data warehouse, while the other two efforts refer to cross-version queries.

Table 5. Summary table for Sect. 5.3

Works	Problem	Input	Output	Method
[52]	Data translation between versions of DW	History of DW data	A derivation of the data of previous version	Transformation matrices that are mappings between the different versions of the DW
[50, 51]	Data translation between versions of DW	History of DW data; Multi-period query	A derivation of the data answering the multi-period query	Data mining techniques that identify DW schema changes and dimension changes, using a normalized matrix
[53]	Data translation between versions of DW; Cross-version queries	History of DW schema	Mapping of previous schemata and data to current schema	Graphs with a simple algebra that describes schema changes and augmented schemata to translate the data from old schemata to current
[49]	Cross version queries & changes of external data sources	History of DW schema; Data providers; Cross version query	Answer to the cross version query	Decompose a query to queries that are correct at each schema version. For the evolution of sources, wrappers notify monitors that activate rules that respond to the change

6 Prospects for Future Research

Handling data and software evolution seems to be a meta-problem that generates problems in specific subareas of computer science and data management. As such, we forecast that research problems around the evolution of data and their structure will never cease to exist.

We have covered the area of logical schema evolution in relational settings, and data warehouses in particular. The evolution of data at the instance level and at the evolution of the schema at the physical level has not been covered in this paper, although both are of great importance.

We also believe that as particular areas of data management have provided ground for research on the problem of evolution in the past (e.g., Conceptual Modeling, XML, Object-Oriented databases, etc.), the future will include research efforts in the hot topics of the day, at any given time period. For example, nowadays, we anticipate that schema-less data, or data with very flexible structures (graphs, texts, JSON objects, etc.) will offer ground for research on the management of their evolution.

Concerning the area of the impact of evolution to ecosystems, the two main areas that seem to require further investigation are: (a) the identification of the constructs that are most sensitive to evolution – ideally via metrics that assess their sensitivity to evolution, and (b) the full automation of the reaction to changes by mechanisms like self-monitoring and self-repairing.

We close with the remark that due to the huge importance and impact of evolution in the lifecycle of both data and software, the potential benefits outweight the (quite significant) risk of pursuing research of both pure scientific nature, in order to find laws and patterns of evolution, and of practical nature, via tools and methods that reduce the pain of evolution's impact.

References

1. Roddick, J.F.: A survey of schema versioning issues for database systems. Inf. Softw. Technol. **37**(7), 383–393 (1995)
2. Hartung, M., Terwilliger, J.F., Rahm, E.: Recent advances in schema and ontology evolution. In: Bellahsene, Z., Bonifati, A., Rahm, E. (eds.) Schema Matching and Mapping. data-centric systems and applications, pp. 149–190. Springer, Heidelberg (2011)
3. Sjøberg, D.: Quantifying schema evolution. Inf. Softw. Technol. **35**(1), 35–44 (1993)
4. Curino, C., Moon, H.J., Tanca, L., Zaniolo, C.: Schema evolution in wikipedia: toward a web information system Benchmark. In: Proceedings of 10th International Conference on Enterprise Information Systems (ICEIS) (2008)
5. Curino, C.A., Moon, H.J., Zaniolo, C.: Graceful database schema evolution: the PRISM workbench. Proc. VLDB Endowment **1**, 761–772 (2008)
6. Curino, C., Moon, H.J., Deutsch, A., Zaniolo, C.: Automating the database schema evolution process. VLDB J. **22**(1), 73–98 (2013)
7. Lin, D.Y., Neamtiu, I.: Collateral evolution of applications and databases. In: Proceedings of the Joint International and Annual ERCIM Workshops on Principles of Software Evolution and Software Evolution Workshops (IWPSE), pp. 31–40 (2009)

8. Wu, S., Neamtiu, I.: Schema evolution analysis for embedded databases. In: Proceedings of the 27th IEEE International Conference on Data Engineering Workshops (ICDEW), pp. 151–156 (2011)
9. Qiu, D., Li, B., Su, Z.: An empirical analysis of the co-evolution of schema and code in database applications. In: Proceedings of the 9th Joint Meeting of the European Software Engineering Conference and the ACM SIGSOFT Symposium on the Foundations of Software Engineering (ESEC/FSE), pp. 125–135(2013)
10. Skoulis, I., Vassiliadis, P., Zarras, A.: Open-source databases: within, outside, or beyond Lehman's laws of software evolution? In: Jarke, M., Mylopoulos, J., Quix, C., Rolland, C., Manolopoulos, Y., Mouratidis, H., Horkoff, J. (eds.) CAiSE 2014. LNCS, vol. 8484, pp. 379–393. Springer, Heidelberg (2014)
11. Skoulis, I., Vassiliadis, P., Zarras, A.: Growing Up with Stability: how Open-Source Relational Databases Evolve. Information Systems in press (2015)
12. Vassiliadis, P., Zarras, A.V., Skoulis, I.: How is life for a table in an evolving relational schema? birth, death and everything in between. In: Johannesson, P., Lee, M.L., Liddle, S.W., Opdahl, A.L., Pastor López, Ó. (eds.) ER 2015. LNCS, vol. 9381, pp. 453–466. Springer, Heidelberg (2015). doi:10.1007/978-3-319-25264-3_34
13. Lehman, M.M., Fernandez-Ramil, J.C.: Rules and tools for software evolution planning and management. In: Software Evolution and Feedback: Theory and Practice. Wiley (2006)
14. Belady, L.A., Lehman, M.M.: A model of large program development. IBM Syst. J. **15**(3), 225–252 (1976)
15. Herraiz, I., Rodriguez, D., Robles, G., Gonzalez-Barahona, J.M.: The evolution of the laws of software evolution: a discussion based on a systematic literature review. ACM Comput. Surv. **46**(2), 1–28 (2013)
16. Lehman, M.M., Fernandez-Ramil, J.C., Wernick, P., Perry, D.E., Turski, W.M.: Metrics and laws of software evolution - the nineties view. In: Proceedings of the 4th IEEE International Software Metrics Symposium (METRICS), pp. 20–34 (1997)
17. Oracle: Oracle Change Management Pack (2014). http://docs.oracle.com/html/A96679_01/overview.htm
18. IBM: Schema changes (2014). http://pic.dhe.ibm.com/infocenter/db2luw/v10r1/index.jsp?topic=%2Fcom.ibm.db2.luw.admin.dbobj.doc%2Fdoc%2Fc0060234.html
19. IBM: IBM DB2 object comparison tool for Z/OS version 10 release 1 (2012). http://www-01.ibm.com/support/knowledgecenter/SSAUVH_10.1.0/com.ibm.db2tools.gou10.doc.ug/gocugj13.pdf?lang=en
20. Microsoft: SQL management studio for SQL server user's manual (2012). http://www.sqlmanager.net/download/msstudio/doc/msstudio.pdf
21. Snaidero, B.: Capture SQL Server Schema Changes Using the Default Trace. Technical report, MSSQLTips (2015). https://www.mssqltips.com/sqlservertip/4057/capture-sql-server-schema-changes-using-the-default-trace/
22. Microsoft: Microsoft SQL server data tools: Database development zero to sixty (2012). http://channel9.msdn.com/Events/TechEd/Europe/2012/DBI311
23. Terwilliger, J.F., Bernstein, P.A., Unnithan, A.: Worry-free database upgrades: automated model-driven evolution of schemas and complex mappings. In: Elmagarmid, A.K., Agrawal, D. (eds.) Proceedings of the ACM SIGMOD International Conference on Management of Data, SIGMOD 2010, Indianapolis, Indiana, USA, 6–10 June 2010, pp. 1191–1194. ACM (2010)
24. Foundation, D.S.: Django (2015). https://www.djangoproject.com/
25. community, S.: South (2015). http://south.readthedocs.org/en/latest/index.html

26. DAINTINESS-Group: Hecate (2015). https://github.com/DAINTINESS-Group/Hecate

27. DAINTINESS-Group: Hecataeus (2015). http://cs.uoi.gr/vassil/projects/hecataeus/index.html

28. Maule, A., Emmerich, W., Rosenblum, D.S.: Impact analysis of database schema changes. In: Schäfer, W., Dwyer, M.B., Gruhn, V. (eds.) ICSE, pp. 451–460. ACM (2008)

29. Papastefanatos, G., Vassiliadis, P., Simitsis, A., Vassiliou, Y.: Policy-regulated management of ETL evolution. J. Data Semant. **13**, 147–177 (2009)

30. Papastefanatos, G., Vassiliadis, P., Simitsis, A., Aggistalis, K., Pechlivani, F., Vassiliou, Y.: Language extensions for the automation of database schema evolution. In: Cordeiro, J., Filipe, J. (eds.) ICEIS (1), pp. 74–81 (2008)

31. Papastefanatos, G., Vassiliadis, P., Simitsis, A., Vassiliou, Y.: Design metrics for data warehouse evolution. In: Li, Q., Spaccapietra, S., Yu, E., Olivé, A. (eds.) ER 2008. LNCS, vol. 5231, pp. 440–454. Springer, Heidelberg (2008)

32. Papastefanatos, G., Vassiliadis, P., Simitsis, A., Vassiliou, Y.: HECATAEUS: regulating schema evolution. In: Proceedings of the 26th IEEE International Conference on Data Engineering (ICDE), pp. 1181–1184 (2010)

33. Papastefanatos, G., Vassiliadis, P., Simitsis, A., Vassiliou, Y.: Metrics for the prediction of evolution impact in ETL ecosystems: a case study. J. Data Semant. **1**(2), 75–97 (2012)

34. Manousis, P., Vassiliadis, P., Papastefanatos, G.: Automating the adaptation of evolving data-intensive ecosystems. In: Ng, W., Storey, V.C., Trujillo, J.C. (eds.) ER 2013. LNCS, vol. 8217, pp. 182–196. Springer, Heidelberg (2013)

35. Manousis, P., Vassiliadis, P., Papastefanatos, G.: Impact analysis and policy-conforming rewriting of evolving data-intensive ecosystems. J. Data Semant. **4**(4), 231–267 (2015)

36. Curino, C., Moon, H.J., Deutsch, A., Zaniolo, C.: Update rewriting and integrity constraint maintenance in a schema evolution support system: PRISM++. PVLDB **4**(2), 117–128 (2010)

37. Mohania, M.: Avoiding re-computation: view adaptation in data warehouses. In: Proceedings of 8th International Database Workshop, Hong Kong, pp. 151–165 (1997)

38. Gupta, A., Mumick, I.S., Rao, J., Ross, K.A.: Adapting materialized views after redefinitions: techniques and a performance study. Inf. Syst. **26**(5), 323–362 (2001)

39. Nica, A., Lee, A.J., Rundensteiner, E.A.: The CVS algorithm for view synchronization in evolvable large-scale information systems. In: Schek, H.-J., Saltor, F., Ramos, I., Alonso, G. (eds.) EDBT 1998. LNCS, vol. 1377, pp. 357–373. Springer, Heidelberg (1998)

40. Golfarelli, M., Rizzi, S.: A survey on temporal data warehousing. IJDWM **5**(1), 1–17 (2009)

41. Wrembel, R.: A survey of managing the evolution of data warehouses. IJDWM **5**(2), 24–56 (2009)

42. Bellahsène, Z.: View adaptation in data warehousing systems. In: Quirchmayr, G., Bench-Capon, T.J.M., Schweighofer, E. (eds.) DEXA 1998. LNCS, vol. 1460, pp. 300–309. Springer, Heidelberg (1998)

43. Bellahsene, Z.: Schema evolution in data warehouses. Knowl. Inf. Syst. **4**(3), 283–304 (2002)

44. Quix, C.: Repository support for data warehouse evolution. In: Gatziu, S., Jeusfeld, M.A., Staudt, M., Vassiliou, Y. (eds.) DMDW. CEUR Workshop Proceedings, vol. 19. CEUR-WS.org (1999)

45. Blaschka, M., Sapia, C., Höfling, G.: On schema evolution in multidimensional databases. In: Mohania, M., Tjoa, A.M. (eds.) DaWaK 1999. LNCS, vol. 1676, pp. 153–164. Springer, Heidelberg (1999)
46. Hurtado, C.A., Mendelzon, A.O., Vaisman, A.A.: Maintaining data cubes under dimension updates. In: Kitsuregawa, M., Papazoglou, M.P., Pu, C. (eds.) ICDE, pp. 346–355. IEEE Computer Society (1999)
47. Hurtado, C.A., Mendelzon, A.O., Vaisman, A.A.: Updating OLAP dimensions. In: Song, I.Y., Teorey, T.J. (eds.) DOLAP, pp. 60–66. ACM (1999)
48. Kaas, C., Pedersen, T.B., Rasmussen, B.: Schema evolution for stars and snowflakes. In: ICEIS (1), pp. 425–433(2004)
49. Wrembel, R., Bebel, B.: Metadata management in a multiversion data warehouse. J. Data Semant. **8**, 118–157 (2007)
50. Eder, J., Koncilia, C., Mitsche, D.: Automatic detection of structural changes in data warehouses. In: Kambayashi, Y., Mohania, M., Wöß, W. (eds.) DaWaK 2003. LNCS, vol. 2737, pp. 119–128. Springer, Heidelberg (2003)
51. Eder, J., Koncilia, C., Mitsche, D.: Analysing slices of data warehouses to detect structural modifications. In: Persson, A., Stirna, J. (eds.) CAiSE 2004. LNCS, vol. 3084, pp. 492–505. Springer, Heidelberg (2004)
52. Eder, J., Koncilia, C.: Changes of dimension data in temporal data warehouses. In: Kambayashi, Y., Winiwarter, W., Arikawa, M. (eds.) DaWaK 2001. LNCS, vol. 2114, pp. 284–293. Springer, Heidelberg (2001)
53. Golfarelli, M., Lechtenbörger, J., Rizzi, S., Vossen, G.: Schema versioning in data warehouses: enabling cross-version querying via schema augmentation. Data Knowl. Eng. **59**(2), 435–459 (2006)

Publishing OLAP Cubes on the Semantic Web

Alejandro Vaisman[(✉)]

Instituto Tecnológico de Buenos Aires, Buenos Aires, Argentina
avaisman@itba.edu.ar

Abstract. The availability of large repositories of semantically annotated data on the web is opening new opportunities for enhancing
Decision-Support Systems. In addition, the advent of initiatives such
as Open Data and Open Government, together with the Linked Open
Data paradigm, are promoting publication and sharing of multidimensional data (MD) on the web. In this paper we address the problem of
representing MD data using Semantic Web (SW) standards. We discuss
how MD data can be represented and queried directly over the SW,
without the need to download data sets into local data warehouses. We
first comment on the RDF Data Cube Vocabulary (QB), the current
W3C recommendation, and show that it is not enough to appropriately
represent and query MD data on the web. In order to be able to support useful Online Analytical Process (OLAP) analysis, extension to QB,
denoted QB4OLAP, has been proposed. We provide an in-depth comparison between these two proposals, and show that extending QB with
QB4OLAP can be done without re-writing the observations, (the largest
part of a QB data set). We provide extensive examples of the QB4OLAP
representation, using a portion of the Eurostat data set and the well-
known Northwind database. Finally, we present a high-level query language, called QL, that allows OLAP users not familiar with SW concepts
or languages, to write and execute OLAP operators without any knowledge of RDF or SPARQL, the standard data model and query language,
respectively, for the SW. QL queries are automatically translated into
SPARQL (using the QB4OLAP metadata) and executed over an endpoint.

Keywords: Data warehousing · OLAP · Semantic web · RDF ·
SPARQL · Linked data

1 Introduction

Data Warehouses (DW) integrate data from multiple sources for analysis and
decision support. They represent data according to *dimensions* and *facts*. The
former reflect the perspectives from which data are viewed. The latter correspond to (usually) quantitative data (also known as *measures*) associated with
different dimensions. Facts can be aggregated and disaggregated through operations called Roll-up and Drill-down, respectively, filtered, by means of Slice and
Dice operations, and so on. This process is called Online Analytical Processing

© Springer International Publishing Switzerland 2016
E. Zimányi and A. Abelló (Eds.): eBISS 2015, LNBIP 253, pp. 32–61, 2016.
DOI: 10.1007/978-3-319-39243-1_2

(OLAP). As an illustration, the facts related to the sales of a company may be associated with the dimensions Time and Location, representing the sales at certain locations, at certain periods of time. Dimensions are modeled as hierarchies of elements (also called members), such that each element belongs to a category (or level) in a hierarchy. DWs and OLAP systems are based on the multidimensional (MD) model, which views data in an n-dimensional space, usually called a data cube, whose axes are the dimensions, and whose cells contain the values for the measures. In the former example, a point in this space could be (*January 2015, Buenos Aires*), where the measure in the cell indicates the amount of the sales in January 2015, at the Buenos Aires branch.

Historically, DW and OLAP had been used as techniques for data analysis, typically using commercial tools with proprietary formats. However, initiatives like Open Data[1] and Open Government[2] are pushing organizations to publish MD data using standards and non-proprietary formats. In the last decade, several open source platforms for Business Intelligence (BI) have emerged, but, at the time this tutorial paper is being written, an open format to publish and share cubes among organizations is still needed. Further, Linked Data [1], a data publication paradigm, promotes sharing and reusing data on the web using Semantic Web (SW) standards and domain ontologies expressed in RDF (the basic data representation layer for the SW) [2], or in languages built on top of RDF (like RDF-Schema [3]). All of the above has widened the spectrum of users, and nowadays, in addition to the typical OLAP analysts, non-technical people are willing to analyze MD data.

1.1 Problem Statement

Two main approaches are found concerning OLAP analysis of MD data on the SW. The first one aims at extracting MD data from the Web, and loading them into traditional data management systems for OLAP analysis. The second one proposes to carry out OLAP-like analysis directly over SW data, typically, over MD data represented in RDF. In this tutorial we focus on the latter approach although, for completeness, in Sect. 2 we discuss and compare both lines of work.

Publishing and analyzing OLAP data directly over the SW, supports the concepts of self-service BI, or on-demand BI, aimed at incorporating web data into the decision-making process with little or no intervention of programmers or designers [4]. Statistical data sets are usually published using the RDF Data Cube Vocabulary (also denoted QB) [5], a W3C recommendation since January, 2014. However, as we explain later, among other limitations, the QB vocabulary does not support the representation of dimension hierarchies and aggregation functions needed for OLAP analysis. To address this challenge, a new vocabulary, called QB4OLAP has been proposed [6]. A key feature of QB4OLAP is that it allows reusing data already published in QB, by means of the addition of the hierarchical structure of the dimensions (and the corresponding instances

---

[1] http://okfn.org/opendata/.

[2] http://opengovdata.org/.

that populate the dimension levels). Once a data cube becomes published using QB4OLAP, users can perform OLAP operations over it. Moreover, high-level languages can be used to seamlessly query these data cubes, as we will show later.

1.2 Running Example

As our running example, we will use statistical data about asylum applications to the European Union (EU), provided by Eurostat, the EU's statistical office[3]. This data set contains information about the number of asylum applicants by month, age, sex, citizenship, application type, and country that receives the application, and it is published using QB in the Eurostat - Linked Data dataspace[4]. For this tutorial, we extended the original QB data cube with dimension hierarchies, as shown in Fig. 1, using the MultiDim conceptual model [7]. The Asylum_applications fact contains only one measure (#applications) that represents the number of applications. This measure can be analyzed according to six analysis dimensions: the sex of the applicant, age which organizes applicants according to their age group, the time of the application (which includes a two-level hierarchy (with levels month and year), the application_type, which tells if the person is a first time applicant or a repeated applicant, and a geographical dimension that organizes countries into continents (the Geography hierarchy), or according to its government type (the Government hierarchy). This geographical dimension participates in the cube with two different roles: The citizenship of the asylum seeker, and the destination country of its application. Usually, these kinds of dimensions are denoted *role-playing* dimensions.

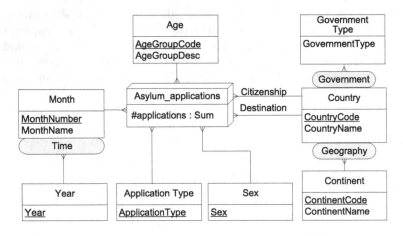

Fig. 1. Conceptual schema of the asylum applications cube

---

[3] http://epp.eurostat.ec.europa.eu/cache/ITY_SDDS/EN/migr_asyapp_esms.htm.
[4] http://eurostat.linked-statistics.org/.

1.3   Contributions

Although QB provides basic multidimensional information, this information is not enough to satisfy OLAP users' needs. In this way, a great part of the benefit of having MD data on the web gets lost. Further, since the QB model does not provide enough information for OLAP analysis, once downloaded, the data must be extended with the typical MD constructs. QB4OLAP has been proposed to address these drawbacks, allowing data owners to publish MD data on the SW, and to enrich existing data sets with structural metadata, and dimensional data. This enrichment can be done semi-automatically (a problem which is beyond the scope of the present paper, and is explained in detail in [8,9]). Also, QB4OLAP data cubes can be created from scratch, for example, integrating on-the-fly, data on the web. Last, but not least, a cube representation like the one allowed by QB4OLAP can not only be used to perform OLAP analysis through queries written in SPARQL [10] (the standard query language for RDF), but to express these queries using a high-level declarative query language, which can be then automatically translated into SPARQL (with the help of the QB4OLAP metadata), allowing non-technical users to perform OLAP data analysis without the need to understand how data are represented. In other words, typical OLAP users could be able to query MD data represented in RDF without the need of having any knowledge of SPARQL.

Concretely, in this tutorial paper we present:

- A comparison between the QB and QB4OLAP vocabularies;
- A description of how QB cubes can be enriched with OLAP metadata and data, and how existing DW can be published using the QB4OLAP vocabulary;
- A user-centric high-level query language, called QL, that expresses the most common OLAP operators independently of the underlying data representation, and a mechanism to automatically translate a QL expression into SPARQL, to query QB4OLAP cubes.

The remainder of the paper is organized as follows. Section 2 discusses related work. In Sect. 3 we introduce the basic concepts used throughout this paper. Section 4 studies the QB vocabulary, and discusses its limitations for representation and querying of MD data. Section 5 presents the QB4OLAP vocabulary, and an in-depth comparison against QB. Section 6 studies the Cube Algebra language, a high-level language to query cubes, and Sect. 7, the translation of Cube Algebra into SPARQL, to query cubes whose underlying representation is based on RDF and the QB4OLAP vocabulary. We conclude in Sect. 8.

This tutorial paper follows the presentation given by the author in the EBISS 2015 Summer School. It is not aimed at presenting original research material, but to put together, in a tutorial style, the main contributions of the work performed by the author in collaboration with other colleagues [6,8,9,11,12].

2   Related Work

As mentioned above, two main approaches concerning OLAP analysis of MD data on the SW can be found in the literature. The first one consists in extracting

MD data from the SW and loading them into traditional MD data management systems for OLAP analysis, while the second one promotes performing OLAP-like analysis directly over SW data.

Along the first line of research, we find the works by Nebot and Llavori [13] and Kämpgen and Harth [14]. The former proposes a semi-automatic method for on-demand extraction of semantic data into an MD database, so data could be analyzed using traditional OLAP techniques. The authors present a methodology for discovering facts in SW data (represented as an OWL[5] ontology), and populating an MD model with such facts. In this methodology, an MD schema is initially designed, indicating the subject of analysis that corresponds to a concept of the ontology, the potential dimensions, and the facts. Then, the dimension hierarchies are created, based on the knowledge available in the domain ontologies (i.e., the inferred taxonomic relationships). Finally, the user specifies the MD queries over the DW. Once queries are executed, a cube is built, and typical OLAP operations can be applied over this cube.

Kämpgen and Harth [14] study the extraction of statistical data published using the QB vocabulary into an MD database. The authors propose a mapping between the concepts in QB, and an MD data model, and implement these mappings via SPARQL queries. In this methodology, the user first defines relevant data sets, which are retrieved from the web, and stored in a local triple store. A relational representation of the MD data model is then created and populated. Over this model, OLAP operations can be performed.

These two efforts are based on traditional MD data management systems, and require the existence of a local DW to store SW data. Also, they do not consider the possibility of directly querying *à la* OLAP MD data over the SW. Thus, a second line of research tries to overcome these drawbacks, exploring data models and tools that allow publishing and performing OLAP-like analysis directly over SW MD. The work we discuss in the remainder, follows this approach.

Terms like self-service BI [4], and Situational BI [15], refer to the capability of incorporating situational data into the decision process with little or no intervention of programmers or designers. In [4], the authors present a framework to support self-service BI, based on the notion of fusion cubes, i.e., multidimensional cubes that can be dynamically extended both in their schema and their instances, and in which data and metadata can be associated with quality and provenance annotations. These frameworks motivate the need for models and tools that allow to query MD data directly over the SW.

The RDF Data Cube vocabulary [5] is aimed at representing, using RDF, statistical data according to the SDMX[6] information model discussed in Sect. 3.2. Although similar to traditional MD data models, the SDMX semantics imposes restrictions on what can be represented using QB. In particular, dimension hierarchies, a key concept in OLAP operations, are not appropriately supported in QB. To overcome this limitation, Etcheverry and Vaisman [6] proposed QB4OLAP, an extension to QB that allows representing analytical data

---

[5] http://www.w3.org/TR/owl2-overview/.
[6] http://SDMX.org.

according to traditional MD models, also proposing a preliminary implementation of some OLAP operators, using SPARQL queries over data cubes specified using QB4OLAP.

In [16] the authors present a framework for performing exploratory OLAP over Linked Open Data sources, where the multidimensional schema of the data cube is expressed in QB4OLAP and VoID. Based on this multidimensional schema the system is able to query data sources, extract and aggregate data, and build an OLAP cube. The multidimensional information retrieved from external sources is also stored using QB4OLAP.

The QB and QB4OLAP approaches will be compared in depth in Sect. 4, and, after this, the paper will be devoted to the study of QB4OLAP and its applications.

3 Preliminary Concepts

In this section we introduce the concepts that we will use in the rest of the paper. To set up a common analysis framework, we first need to briefly define the MD model for OLAP that will be used in our study. We do this in the first part of the section. In the second part we discuss statistical databases (SDB), and introduce the SDMX model, on which QB is based. We conclude with a definition of the basic SW concepts that we will need in the sequel.

3.1 OLAP

A broad number of MD models can be found in the literature [17–19]. We now describe the MD model for OLAP that we will use in our study.

In OLAP, data are organized as *hypercubes* whose axes are called *dimensions*. Each point in this MD space is mapped into one or more spaces of *measures*, representing *facts* that are analyzed along the cube's dimensions. Dimensions are structured in *hierarchies* that allow analysis at different aggregation *levels*. The actual values in a dimension level are called *members*.

A *Dimension Schema* is composed of a non-empty finite set of levels, with a distinguished level denoted *All*. We denote '→' a partial order on these levels; the reflexive and transitive closure of '→' ('→*') has a unique bottom level and a unique top level (the latter denoted *All*). Levels can have attributes describing them. A *Dimension Instance* assigns to each dimension level in the dimension schema a set of dimension *members*. For each pair of levels (l_j, l_k) in the dimension schema, such that $l_j \rightarrow l_k$, a relation (denoted *rollup*) is defined, associating members from level l_j with members of level l_k. Although in practice, most MD models assume a function between the instances of parent and child dimension levels, we support relations between them, meaning that each member in the child level many have more than one associated member in the parent level, and vice versa (hierarchies including rollup relations are called non-strict). Cardinality constraints on these relations are then used to restrict the number of level members related to each other [7]. A *Cube Schema* contains a set of dimension schemas and a set of measures, where for each measure an aggregate function is specified. A *Cube Instance*, corresponding to a cube schema, is a partial function mapping coordinates from dimension instances into measure values.

A well-known set of operations is defined over cubes. For instance, based on the algebra sketched in [20], the *Roll-Up* operation summarizes data in a cube, along a dimension hierarchy. Analogously, *Drill-Down* disaggregates previously summarized data, and can be considered the inverse of Roll-Up. The *Slice* operation drops a dimension from a cube. The *Dice* operation receives a cube $\mathcal{C}$, and a first order formula ϕ over levels and measures in $\mathcal{C}$, and returns a new cube with the same schema, and whose instances are the ones that satisfy ϕ. There are more complex operators, but for the sake of simplicity, we will limit ourselves to the ones mentioned above.

3.2   Statistical Databases and the SDMX Model

Statistical Data Bases (SDB) also organize data as hypercubes whose axes are dimensions. Each point in this multidimensional space is mapped through observations into one or more spaces of measures. Dimensions are structured in classification hierarchies that allow analysis at different levels of aggregation. The Statistical Data and Metadata eXchange initiative (SDMX) proposes several standards for the publication, exchange and processing of statistical data, and defines an information model [21] from which we summarize some concepts next, since QB is based on SDMX.

In the SDMX model, a *Dimension* denotes a metadata concept used to classify a statistical series, e.g., a statistical concept indicating a certain economic activity. Two particular dimensions are identified: *TimeDimension*, specifying a concept used to convey the time period of the observation in a data set; and *MeasureDimension*, whose purpose is to specify formally the meaning of the measures and to enable multiple measures to be defined and reported in a data set. A *Primary Measure* denotes a metadata concept that represents the phenomenon to be measured in a data set. Dimensions, measures, and attributes are called *Components*.

Codelists enumerate a set of values to be used in the representation of dimensions, attributes, and other structural parts of SDMX. Additional structural metadata can indicate how codes are organized into hierarchies. Through the inheritance abstraction mechanism, the codelist comprises one or more codes, and the code itself can have one or more children codes in the (inherited) hierarchy association. Note that a child code can have only one parent code in this association.

A *Data Set* denotes a set of observations that share the same dimensionality, which is specified by a set of unique components (e.g., dimensions, measures). Each data set is associated with structural metadata, called *Data Structure Definition* (DSD), that includes information about how concepts are associated with the measures and dimensions of a data cube along descriptive (structural) metadata.

The value of the variable being measured for the concept associated to the *PrimaryMeasure* in the DSD is called an *Observation*. Each observation associates an *observation value* with a *key value*.

Several operators are defined over SDBs, although the SDMX standard does not define operators over data sets. Instead, it provides a mechanism to restrict the values within a data set via *constraints*. For example, the *CubeRegions* constraint, allows specifying a set of component values, defining a subset of the total range of the content of a data structure. The application of this constraint results in a *slice* of the original data set, fixing values for some components (e.g.: selecting some years in a TimeDimension). Therefore, the name *slice* may be misleading for OLAP practitioners, since in OLAP, a slicing operation reduces the cube's dimensionality, as explained in Sect. 3.1.

3.3   RDF and the Semantic Web

The Resource Description Framework (RDF) is a data model for expressing assertions over resources identified by an internationalized resource identifier (IRI). Assertions are expressed as triples of the form $(subject, predicate, object)$. A set of RDF triples or *RDF data set* can be seen as a directed graph where *subject* and *object* are nodes, and *predicates* are arcs. Data values in RDF are called *literals*. *Blank nodes* are used to represent anonymous resources or resources without an IRI, typically with a structural function, e.g., to group a set of statements. Subjects must always be resources or blank nodes, predicates are always resources, and objects could be resources, blank nodes or literals. A set of reserved words defined in RDF Schema (called the rdfs-vocabulary)[3] is used to define classes, properties, and to represent hierarchical relationships between them. For example, the triple $(s,$ rdf:type, $c)$ explicitly states that s is an instance of c but it also implicitly states that object c is an instance of rdf:Class since there exists at least one resource that is an instance of c. Many formats for RDF serialization exist. In this paper we use Turtle [22].

SPARQL 1.1 [10] is the W3C standard query language for RDF, at the time this paper is being written. The query evaluation mechanism of SPARQL is based on subgraph matching: RDF triples are interpreted as nodes and edges of directed graphs, and the query graph is matched to the data graph, instantiating the variables in the query graph definition. The selection criteria is expressed as a graph pattern in the WHERE clause of a SPARQL query. Relevant to OLAP queries, SPARQL supports aggregate functions and the GROUP BY clause, as in classic SQL.

Due to space limitations, in the remainder we assume the reader is familiar with the basic notions of RDF and SPARQL.

4   QB: The RDF Data Cube Vocabulary

We now study in detail the QB vocabulary, and discuss its possibilities and limitations for representing and analyzing MD data.

4.1 Vocabulary Description

As mentioned above, QB is the W3C recommendation to publish statistical data and metadata in RDF, following the Linked Data principles. QB is based on the SDMX Information Model described in Sect. 3.2, and is the evolution of two previous attempts to represent statistical data in RDF: the Statistical Core Vocabulary (SCOVO) [23], and SDMX-RDF [24]. Figure 2 (taken from the W3C recommendation document [5]) depicts the QB vocabulary. Capitalized terms represent RDF classes and non-capitalized terms represent RDF properties. An arrow from class A to class B, labeled rel means that rel is an RDF property with domain A and range B. White triangles represent sub-classes or sub-properties. We describe the QB vocabulary next.

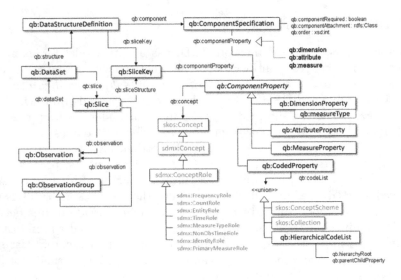

Fig. 2. The QB vocabulary (cf. [5])

The schema of a data set is specified by means of the DSD (like in SDMX), an instance of the class qb:DataStructureDefinition. This specification comprises a set of *Component* properties, instances of the class qb:ComponentProperty (in italics in Fig. 2), representing *Dimensions*, *Measures*, and *Attributes*. This is shown in Example 1. Note that a DSD can be shared by many data sets by means of the qb:structure property. *Observations* (in OLAP terminology, *facts*), are instances of the class qb:Observation, and represent points in an MD data space indexed by dimensions. They are associated with data sets (instances of the class qb:DataSet), through the qb:dataSet property (see Example 2). Each observation can be linked to a value in each dimension of the DSD via instances of qb:DimensionProperty; analogously, values for each observation are associated with measures via instances of the class qb:MeasureProperty. Instances of the class qb:AttributeProperty are used to associate attributes with observations.

Finally, note that QB allows observations in a data set to be expressed at different levels of granularity in each dimension. For example, one observation may refer to a country, and another one may refer to a region.

Component properties are not directly related to the DSD: the class qb:ComponentSpecification is an intermediate class which allows to specify additional attributes for a component in a DSD. For example, a component may be tagged as *required* (i.e., mandatory), using the qb:componentRequired property. Components that belong to a specification are linked using specific properties that depend on the type of the component, that is, qb:dimension for dimensions, qb:measure for measures, and qb:attribute for attributes. Component specifications are linked to DSDs via the qb:component property. For instance, in Example 1 we can see how dimensions are defined in the DSD, through the qb:dimension and qb:component properties.

In order to allow reusing the concepts defined in the SDMX Content Oriented Guidelines [25], QB provides the qb:concept property which links components to the general concepts they represent. The latter are modeled using the skos:Concept class defined in the SKOS vocabulary.[7]

Although QB can define the structure of a fact (via the DSD), it does not provide a mechanism to represent an OLAP dimension structure (i.e., the dimension levels and the relationships between levels). However, QB allows representing hierarchical relationships between level members in the dimension instances. The QB specification describes three possible scenarios with respect to the organization of dimensions, as we explain next.

- If there is no need to define hierarchical relationships within dimension members, QB recommends representing the members using instances of the class skos:Concept, and the set of admissible values using skos:ConceptScheme. A SKOS concept scheme allows organizing one or more SKOS concepts, linked to the concept schemes they belong to, via the skos:inScheme property.
- To represent hierarchical relationships, the recommendation is to use the semantic relationship skos:narrower, with the following meaning: if two concepts A and B are related using skos:narrower, B represents a finer concept than A (e.g., animals skos:narrower mammals). In addition, SKOS defines a skos:hasTopConcept property, which allows linking a concept scheme to the (possibly many) most general concept it contains. To reuse existing data, QB provides the class qb:HierarchicalCodeList. An instance of this class defines a set of root concepts in the hierarchy using qb:hierarchyRoot and a parent-child relationship via qb:parentChildProperty which links a term in the hierarchy to its immediate sub-terms.

Finally, *Slices* represent subsets of observations. They are not defined as operators over an existing cube, but as new structures and new instances (observations), where one or more values of dimension members are fixed. The structure of a slice is defined using a DSD, and an instance of the qb:SliceKey class.

---

[7] http://www.w3.org/TR/2009/REC-skos-reference-20090818/.

Example 1 below presents the triples that represent a portion of the structure of the QB data set in our running example. Note that components are defined as RDF blank nodes.

Example 1 (Data Set Structure Definition).

```
1  @prefix qb: <http://purl.org/linked−data/cube#> .
2  @prefix sdmx−dimension: <http://purl.org/linked−data/sdmx/2009/dimension#> .
3  @prefix sdmx−measure: <http://purl.org/linked−data/sdmx/2009/measure#> .
4  @prefix dsd: <http://eurostat.linked−statistics.org/dsd/> .
5  @prefix property: <http://eurostat.linked−statistics.org/property#> .
6
7  dsd:migr_asyappctzm rdf:type qb:DataStructureDefinition ;
8        qb:component [qb:dimension property:age] ;
9        qb:component [qb:dimension property:geo] ;
10       qb:component [qb:dimension property:sex] ;
11       qb:component [qb:dimension property:citizen] ;
12       qb:component [qb:dimension property:asyl_app] ;
13       qb:component [qb:dimension sdmx−dimension:refPeriod] ;
14       qb:component [qb:measure sdmx−measure:obsValue] .
15
16       <http://eurostat.linked−statistics.org/data/migr_asyappctzm> qb:structure dsd:migr_asyappctzm
```

Line 7 defines the IRI of the DSD. The lines that follow, indicate the components of such structure, and Line 16 tells that the DSD is the structure of the data set in the subject of the triple.   □

Continuing with the Eurostat running example, Example 2 below shows the triples that represent an observation (in OLAP jargon, a *fact*), corresponding to the schema above.

Example 2 (Observations). The following triples represent an observation corresponding to the number of citizens of Andorra submitting applications to migrate to Austria in 2014.

```
1  @prefix data:<http://eurostat.linked−statistics.org>;
2  <http://eurostat.linked−statistics.org/data/migr_asyappctzm#M,AD,F,TOTAL,ASY_APP,AT,2014M10>
3        a qb:Observation ;
4        qb:dataSet <http://data/migr_asyappctzm> ;
5        property:age data:dic/age#TOTAL;
6        property:geo data:dic/geo#AT;
7        property:sex data:dic/sex#F;
8        property:citizen data:dic/citizen#AD;
9        property:asyl_app data:dic/asyl_app#ASY_APP;
10       sdmx−dimension:refPeriod 2014−10−0;
11       sdmx−measure: obsValue 0 .
```

Line 2 tells that the IRI in the subject is an instance of the class qb:Observation, and Line 4 indicates the data set to which the observation belongs. The other triples correspond to the dimension instances and the observed value (the measure, in Line 11).   □

4.2   Is QB Suitable for OLAP?

Although QB can be used to publish MD observations, it cannot represent the most typical features of the MD model that are used to navigate data in an OLAP fashion. We discuss this next.

1. *QB does not provide native support for dimension structures.* Typical OLAP operations, like Roll-up and Drill-down, rely on the organization of dimension members into hierarchies that define aggregation levels. However, as explained above, QB cannot represent the structural metadata needed to appropriately support such operations. The mechanisms described in Sect. 4.1 allows only to organize dimension members hierarchically, that means, they can only represent relationships between instances, for example, to say that Argentina is a finer concept than South America, but not to say that Argentina is a country, South America is a continent, and that countries aggregate over continents.

2. *QB does not provide native support to represent aggregate functions.* Most OLAP operations aggregate or disaggregate cube data along a dimension (e.g., a Roll-up operation over the Time dimension can aggregate measure values from the Month level up to Year level), using an aggregate function defined for each measure. Normally, it is not possible to assume a single aggregate function for all measures. The ability to link each measure with an aggregation function is not present in QB.

3. *QB does not provide native support for descriptive attributes.* In the MD model, each dimension level is associated with a set of *attributes* that describe the characteristics of the dimension members (e.g. the level Country may have the attributes countryName, area, etc.), and one or more *identifiers* [7]. However, in QB, dimension members are represented as coded values, which in most cases are represented as IRIs (although this is not mandatory). We will see later, that this limitation can have impact over some operations, typically, when dicing over a dimension.

5   The QB4OLAP Vocabulary

From the discussion in Sect. 4.2, the need of a more powerful vocabulary was evident. Thus, the QB4OLAP[8] vocabulary has been proposed, extending QB with a set of RDF terms that allow representing the most common features of the MD model. The main features of QB4OLAP are:

– QB4OLAP can represent the most common features of the MD model. Given that there is no standard (or widely accepted) conceptual model for OLAP, the features considered were based on the MultiDim model [7].
– QB4OLAP includes the metadata needed to automatically implement OLAP operations as SPARQL queries. Using these metadata (e.g., the aggregation paths in a dimension), the operations could be written in a high-level language (or submitted using a graphic navigation tool), and translated into SPARQL. In this way, OLAP users, with no knowledge of SPARQL at all, would be able to exploit data on the SW.
– QB4OLAP allows operating over already published observations which conform to DSDs defined in QB, without the need of rewriting the existing observations, and with the minimum possible effort. Note that in a typical MD

---

[8] http://purl.org/qb4olap/cubes.

model, observations are the largest part of the data, while dimensions are usually orders of magnitude smaller.

Figure 3 depicts the QB4OLAP vocabulary. Original QB terms are prefixed with "qb:", and QB4OLAP terms are prefixed with "qb4o:", displayed in gray background. Capitalized terms represent RDF classes, non-capitalized terms represent RDF properties; capitalized terms in italics represent class instances. An arrow from class A to class B, labeled rel means that rel is an RDF property with domain A and range B. White triangles represent sub-class or sub-property relationships. Black diamonds represent rdf:type relationships (instances). We present QB4OLAP distinctive features next.

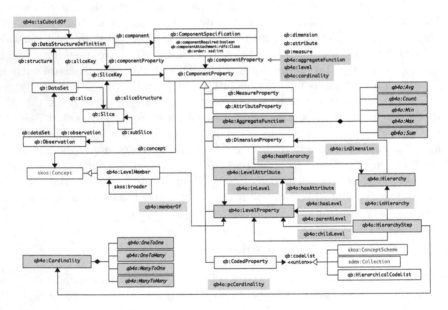

Fig. 3. QB4OLAP vocabulary (cf. [12])

5.1   Dimension Structure in QB4OLAP

As already mentioned, dimension hierarchies and levels are crucial features in an MD model for OLAP. Therefore, QB4OLAP introduced classes and properties to represent them. A key difference between QB and QB4OLAP is that, in the latter, facts represent relationships between dimension levels, and fact instances (observations) map level members to measure values; on the other hand, in QB, observations map dimension members to measure values. In other words, QB4OLAP represents the structure of a data set in terms of dimension levels and measures, instead of dimensions and measures. In QB4OLAP, dimension levels are represented in the same way in which QB represents dimensions: as classes of properties. The class qb4o:LevelProperty represents dimension levels. Since it is declared as a sub-class of qb:ComponentProperty, the schema of the cube can be specified in terms of dimension levels, using the (QB) class qb:DataStructureDefinition (allowing reusing existing QB observations, if needed). To represent aggregate functions the class qb4o:AggregateFunction is defined. The property qb4o:aggregateFunction associates measures with aggregate functions, and, together with the concept of component sets, allows a given measure to be associated with different aggregate functions in different cubes, addressing one of the drawbacks of QB. Finally, when a fact (observation) is related to more than one dimension level member (this is called a many-to-many dimension [7]), the property qb4o:cardinality allows representing the cardinality of this relationship.

Example 3 below, shows how the cube in our Eurostat running example would look like in QB4OLAP. Figure 4 presents the definition of the prefixes that we will use in the sequel.

```
1 PREFIX qb: <http://purl.org/linked-data/cube#>
2 PREFIX qb4o: <http://purl.org/qb4olap/cubes#>
3
4 #QB4OLAP schema and instances
5 PREFIX schema: <http://www.fing.edu.uy/inco/cubes/schemas/migr_asyapp#>;
6 PREFIX instances: <http://www.fing.edu.uy/inco/cubes/instances/migr_asyapp#>
```

Fig. 4. RDF prefixes to be used in the examples

Example 3 (Eurostat Cube Structure in QB4OLAP). Below, we show the structure of a data cube for the Eurostat example, represented using QB4OLAP. The reader is suggested to compare against the DSD in Example 1.

```
1 schema:migr_asyappctzmQB4O rdf:type qb:DataStructureDefinition;
2
3      qb:component [ qb4o:level property:age ; qb4o:cardinality qb4o:ManyToOne ] ;
4      qb:component [ qb4o:level property:geo ; qb4o:cardinality qb4o:ManyToOne ] ;
5      qb:component [ qb4o:level property:sex ; qb4o:cardinality qb4o:ManyToOne ] ;
6      qb:component [ qb4o:level property:citizen qb4o:cardinality qb4o:ManyToOne ] ;
7      qb:component [ qb4o:level property:asyl_app ; qb4o:cardinality qb4o:ManyToOne ] ;
8      qb:component [ qb:measure sdmx-measure:obsValue; qb4o:aggregateFunction qb4o:sum ] ;
9
10 <http://eurostat.linked-statistics.org/data/migr_asyappctzm> qb:structure
11                                                    schema:migr_asyappctzmQB4O.
12 sdmx-measure:obsValue a qb:MeasureProperty;
13                     rdfs:label "Number of applications"@en; rdfs:range xsd:integer .
```

Note that, opposite to QB, the structure is defined in terms of dimension levels, which represent the granularity of the observations in the data set. Each level is associated to a cardinality, using the property qb4o:cardinality. In this case, all cardinalities are many-to-one, indicating that an observation is associated to exactly one member in every dimension level. To avoid rewriting the observations, a QB4OLAP DSD schema:migr_asyappctzmQB4O is created, and associated with the data set <http://eurostat.linked-statistics. org/data/migr_asyappctzm> (recall that in Example 1, the data set structure was dsd:migr_asyappctzm). This allows reusing, as QB4OLAP level properties, the dimension properties already defined in the QB structure, allowing to use the existing observations, since the data set will "point" to this new DSD. Thus, we must declare those properties as instances of qb4o:LevelProperty. For example, for the Time dimension, we must define (we explain this dimension in detail later):

```
1  schema:timeDim a qb:DimensionProperty ;
2      rdfs:label "Time dimension"@en ;
3      qb4o:hasHierarchy schema:timeHier .
4
5  sdmx−dimension:refPeriod a qb4o:LevelProperty ;
6                 rdfs:label "Month level"@en .
```

We can see that sdmx-dimension:refPeriod (the Time dimension) is redefined as a dimension level using the class qb4o:LevelProperty; a dimension schema:timeDim is defined using the QB class qb:DimensionProperty. In addition, a dimension hierarchy schema:timeHier is defined. Since the dimension levels defined in this way are the lowest ones in the dimension hierarchies, a QB4OLAP cube schema can then be defined using these properties. We explain this below.   □

Dimension hierarchies are represented using the class qb4o:Hierarchy; further, the properties qb4o:hasHierarchy and qb4o:inDimension, tell that a dimension contains a certain hierarchy, and that a certain hierarchy belongs to a dimension, respectively. Also, hierarchies are composed of levels, and the relationship between levels in a hierarchy may have different cardinality constraints (e.g. one-to-many, many-to-many, etc.). We call these relationships *hierarchy steps*, which are represented by the class qb4o:HierarchyStep. Each hierarchy step is linked to its two component levels using the qb4o:childLevel and qb4o:parentLevel properties, and can be attached to the hierarchy it belongs to, using the property qb4o:inHierarchy. The property qb4o:pcCardinality represents the cardinality constraints of the relationships between level members in this step, associating a hierarchy with a member of the qb4o:Cardinality class, whose instances are depicted in Fig. 3. Example 4 shows a part of the definition of the dimension hierarchies for our running example.

Example 4 (Dimension Structure and Hierarchies in QB4OLAP). In addition to the definition of the Time dimension structure (schema:timeDim) shown in Example 3, we can define one or more hierarchies, and declare which dimension they belong to, and the levels that they traverse. In this example, we create a hierarchy denoted schema:timeHier, with two levels, sdmx-dimension:refPeriod, and schema:year, representing the aggregation levels month (the bottom level) and year, respectively. Also, the distinguished level All is defined, as schema:timeAll. Below, we show these definitions.

```
1  schema:timeHier a qb4o:Hierarchy ;
2      rdfs:label "Time Hierarchy"@en ;
3      qb4o:inDimension schema:timeDim ;
4      qb4o:hasLevel sdmx−dimension:refPeriod, schema:year , schema:timeAll .
5
6  sdmx−dimension:refPeriod a qb4o:LevelProperty ;
7      rdfs:label "Month level"@en .
8
```

```
9  schema:year a qb4o:LevelProperty ;
10      rdfs:label "Year"@en .
11
12  schema:timeAll a qb4o:LevelProperty ;
13      rdfs:label "All dates"@en .
```

We remark that the lowest granularity level for the time dimension is defined as in QB (i.e., sdmx-dimension:refPeriod), but as a dimension level instead of a dimension.

The parent-child relationships between levels are defined as *hierarchy steps*, using the class qb4o:HierarchyStep, as we show below.

```
1  _:ih21 a qb4o:HierarchyStep ;
2      qb4o:inHierarchy schema:timeHier ;
3      qb4o:childLevel sdmx−dimension:refPeriod ;
4      qb4o:parentLevel schema:year; qb4o:pcCardinality qb4o:ManyToOne .
5
6  _:ih22 a qb4o:HierarchyStep;
7      qb4o:inHierarchy schema:timeHier ;
8      qb4o:childLevel schema:year ;
9      qb4o:parentLevel schema:timeAll ; qb4o:OneToManyToOne .
```

Note that we indicated, for each step (represented using a blank node), to which hierarchy it belongs, which level is the parent (i.e., the level with coarser granularity), and which level is the child (i.e., the level with finer granularity), and the cardinality of the relationship.   □

Finally, in order to address the lack of support for *level attributes* in QB, QB4OLAP provides the class of properties qb4o:LevelAttribute. This class is linked to qb4o:LevelProperty, via the qb4o:hasAttribute property. For completeness, QB4OLAP includes the qb4o:inLevel property, with domain in the class qb4o:LevelAttribute and range in the class qb4o:LevelProperty. The qb4o:inLevel property is rarely used, but is included for completeness, as kind of an "inverse" of qb4o:hasAttribute (note that RDF does not allow to represent the inverse of a property). Level attributes are useful in OLAP to filter cubes according to some attribute property. Example 5 shows the definition of an attribute for the time dimension level sdmx-dimension:refPeriod.

Example 5 (Level Attributes). For this example, assume we add attribute schema: monthNumber to the level sdmx-dimension:refPeriod in the time dimension.

```
1  sdmx−dimension:refPeriod qb4o:hasAttribute schema:monthNumber .
2
3  schema:monthNumber a qb4o:LevelAttribute;
4      rdfs:label "Month number"@en.
```

Note that the attribute schema:monthNumber is declared indicating that it is an instance of the class qb4o:LevelAttribute.   □

5.2   Dimension Instances in QB4OLAP

Typically, instances of OLAP dimensions levels are composed of so-called level members. In QB4OLAP, level members are represented as instances of the class

qb4o:LevelMember, which is a sub-class of skos:Concept. Members are associated with the level they belong to, using the property qb4o:memberOf, whose
semantics is similar to skos:member. Rollup relationships between members are
expressed using the property skos:broader. The choice of this property, instead
of skos:narrower, like it is recommended in QB, aims at indicating that the hierarchies of level members are usually navigated from finer granularity concepts
up to higher granularity concepts. Example 6 below illustrates this.

Example 6 (Dimension Instances in QB4OLAP). We show now some examples
of members of levels in the dimension schema:timeDim.

```
1   @prefix time:<http://purl.org/qb4olap/dimensions/time#> .
2
3   time:TOTAL
4        qb4o:memberOf schema:timeAll .
5
6   time:200801
7        qb4o:memberOf sdmx−dimension:refPeriod;
8        skos:broader time:2008 .
9   ...
10  time:2008
11       qb4o:memberOf schema:year;
12       skos:broader time:TOTAL .
13  ...
14  time:2009
15       qb4o:memberOf schema:year;
16       skos:broader time:TOTAL .
17  ...
18  time:201401
19       qb4o:memberOf sdmx−dimension:refPeriod;
20       skos:broader time:2014 .
21  ...
22  time:2014
23       qb4o:memberOf schema:year;
24       skos:broader time:TOTAL .
25  ...
```

In Lines 6 through 8 we indicate that the month January of 2008 belongs
to level sdmx-dimension:refPeriod, and rolls up to the element time:2008, an IRI
representing the year 2008. In turn, time:2008, defined in Lines 10 through 12,
rolls up to the level time:TOTAL, which represents the distinguished member all
(although it is not mandatory to indicate this element).

Analogously to level members, we must define the instances of level
attributes. For this, associate the IRIs corresponding to level members, with
literals corresponding to attribute values (i.e., attribute instances). In our example, for the Time dimension we have:

```
1   time:201401 schema:monthNumber "201401"^^xsd:integer .
```

Note that, opposite to level members, which are IRIs, attribute instances are
always literals (since QB4OLAP does not define, for attributes, a class analogous
to qb4o:MemberOf).   □

5.3   How Can We Use QB4OLAP?

There are three basic ways of using QB4OLAP: (a) To enrich an existing data
set published in QB, with structural metadata and dimensional data; (b) To

publish an existing data cube/data warehouse; (c) To build a new cube, using QB4OLAP, from scratch. We already discussed option (a). We do not specifically address option (c) here, since it comprises the tasks of the first two ones. We briefly address option (b) in this section.

To illustrate how we can publish an existing DW on the SW using QB4OLAP, we use the well-known Northwind DW (see [7] for a detailed explanation of the Northwind DW design). Figure 5 shows the conceptual model of the Northwind DW using the MultiDim model.

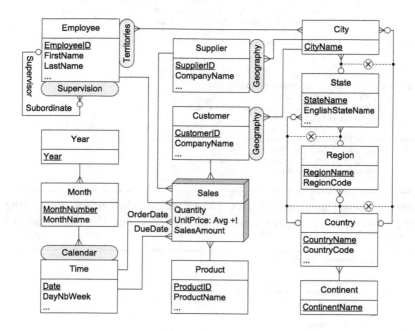

Fig. 5. Conceptual schema of the Northwind DW

It has been already shown that most of the widely used features of the MultiDim conceptual model, and, in general, of the MD model, can be represented using QB4OLAP [12]. Therefore, we do not extend here on this explanation, but below, we give some examples using the Northwind DW.

Example 7 (Northwind DW Structure Definition). This example shows a portion of the DSD that exposes the structure of the Nortwhind DW in QB4OLAP. The DSD is denoted nw:Northwind. It comprises nine dimensions and six measures.

```
1  @prefix nw: <http://dwbook.org/cubes/schemas/northwind#> .
2
3  # Cube definition
4
5  nw:Northwind a qb:DataStructureDefinition ;
6
7  # Lowest level for each dimension in the cube
```

```
 8   qb:component [ qb4o:level nw:employee ; qb4o:cardinality qb4o:ManyToOne ] ;
 9   qb:component [ qb4o:level nw:orderDate ; qb4o:cardinality qb4o:ManyToOne ] ;
10   qb:component [ qb4o:level nw:dueDate ; qb4o:cardinality qb4o:ManyToOne ] ;
11   qb:component [ qb4o:level nw:shippedDate ; qb4o:cardinality qb4o:ManyToOne ] ;
12   qb:component [ qb4o:level nw:product ; qb4o:cardinality qb4o:ManyToOne ] ;
13   qb:component [ qb4o:level nw:order ; qb4o:cardinality qb4o:OneToOne ] ;
14   qb:component [ qb4o:level nw:shipper ; qb4o:cardinality qb4o:ManyToOne ] ;
15   qb:component [ qb4o:level nw:supplier ; qb4o:cardinality qb4o:ManyToOne ] ;
16   qb:component [ qb4o:level nw:customer ; qb4o:cardinality qb4o:ManyToOne ] ;
17
18   # Measures in the cube
19   qb:component [ qb:measure nw:quantity ; qb4o:hasAggregateFunction qb4o:sum ] ;
20   qb:component [ qb:measure nw:unitPrice ; qb4o:hasAggregateFunction qb4o:avg ] ;
21   qb:component [ qb:measure nw:discount ; qb4o:hasAggregateFunction qb4o:avg ] ;
22   qb:component [ qb:measure nw:salesAmount ; qb4o:hasAggregateFunction qb4o:sum ] ;
23   qb:component [ qb:measure nw:freight ; qb4o:hasAggregateFunction qb4o:sum ] ;
24   qb:component [ qb:measure nw:netAmount ; qb4o:hasAggregateFunction qb4o:sum ] .
25
26   # Measures definition
27
28   nw:quantity a rdf:Property , qb:MeasureProperty ;
29           rdfs:label "Quantity"@en ;
30           rdfs:subPropertyOf sdmx−measure:obsValue ;
31           rdfs:range xsd:positiveInteger .
32
33   nw:unitPrice a rdf:Property , qb:MeasureProperty ;
34           rdfs:label "Unit Price"@en ;
35           rdfs:subPropertyOf sdmx−measure:obsValue ;
36           rdfs:range xsd:decimal .
37   ...
```

Next, we show the schema of part of the Employee dimension, illustrating the representation of the recursive Supervision hierarchy, and the definition of level attributes.

```
 1   # −− Employee dimension definition
 2
 3   nw:employeeDim a rdf:Property , qb:DimensionProperty ;
 4           rdfs:label "Employee Dimension"@en ;
 5           qb4o:hasHierarchy nw:supervision , nw:territories .
 6
 7   # −− Supervision hierarchy
 8
 9   nw:supervision a qb4o:Hierarchy ;
10              rdfs:label "Supervision Hierarchy"@en ;
11                  qb4o:inDimension nw:employeeDim ;
12            qb4o:hasLevel nw:employee .
13
14   _:supervision_hs1 a qb4o:HierarchyStep ;
15              qb4o:inHierarchy nw:supervision ;
16              qb4o:childLevel nw:employee ;
17              qb4o:parentLevel nw:employee ;
18              qb4o:pcCardinality qb4o:ManyToOne .
19   ...
20
21   # −− Employee level
22
23   nw:employee a qb4o:LevelProperty ;
24          rdfs:label "Employee Level"@en ;
25          qb4o:hasAttribute nw:employeeID ;
26          qb4o:hasAttribute nw:firstName ;
27          qb4o:hasAttribute nw:lastName ;
28          qb4o:hasAttribute nw:title ;
29          ... .
30
31   nw:employeeID a qb4o:LevelAttribute ;
32              rdfs:label "Employee ID"@en ;
```

```
33              rdfs:range xsd:positiveInteger .
34  nw:firstName a qb4o:LevelAttribute ;
35                 rdfs:label "First Name"@en ; rdfs:range xsd:string .
36  nw:lastName a qb4o:LevelAttribute ;
37                 rdfs:label "Last Name"@en ;
38                 rdfs:range xsd:string .
39  nw:title a qb4o:LevelAttribute ;
40                 rdfs:label "Title"@en ;
41                 rdfs:range xsd:string .
42  ...
```

We can see that, in the recursive hierarchy nw:supervision, there is only one level, nw:employee, that is also the parent and child level of the hierarchy step _:supervision_hs1 (the level All can be omitted). We can also see some of the dimension level attributes, and their definitions.   □

The translation from an existing data cube (for example, a cube represented in the relational model), can be done in an automatic way, using the R2RML standard.[9] The study of this mechanism is out of the scope of this paper (see [26] for an implementation).

In the next section we use the Eurostat data cube to illustrate how we can query it using a high-level language, and its automatic translation into SPARQL.

6   Querying QB4OLAP Cubes

The machinery described above can be applied to query data cubes on the SW, following the approach presented in [20], where a clear separation between the conceptual and the logical levels is made, and a high-level language, called Cube Algebra, is defined. Cube Algebra is a user-centric language operating at the conceptual level. This is the reason why the design of QB4OLAP puts emphasis on representing most of the features of OLAP conceptual models. To take advantage of the vocabulary, a subset of Cube Algebra, called QL, was defined, in a way such that the user can write her queries at the conceptual level, and these queries will be automatically translated into a SPARQL query over the QB4OLAP-based RDF representation (at the logical level). There are also a set of rules to ameliorate and simplify QL queries before obtaining an equivalent SPARQL query, which we explain succinctly below.

Remark 1. The content of this section, is based on the work in [27,28], adapted and simplified for the EBISS 2015 tutorial.

6.1   The QL Language

Ciferri et al. [20] have shown that, opposite to the usual belief, most of the MD data models in the literature operate at the logical level rather than at a conceptual level, and that the data cube is far from being the focus of these models. Therefore, the authors proposed a model and an algebra where the data

---

[9] http://www.w3.org/TR/r2rml/.

cube is a first-class citizen, and OLAP operators are used to manipulate the only type of this model: again, the data cube. Following these ideas, Gómez et al. [17] showed that such a model can be used to seamlessly perform OLAP analysis over discrete and continuous geographic data. That means, the user will write the queries in Cube Algebra, without caring about which kind of data lies underneath. The framework takes care of the spatial data management, and of translating the expressions into the language supported by the underlying database (PostGIS in the case of [17]). Along these lines, the use of a high-level query language (as mentioned, called QL), based on the Cube Algebra, for querying cubes represented in RDF following the QB4OLAP model, has been proposed. In this way, the user will only see a collection of dimensions, dimension levels, and measures, and will write the queries in QL, which will then be translated to SPARQL, and executed on the QB4OLAP underlying data cube.

In this section we briefly outline the portion of QL that we will use in the sequel. We remark that we have simplified the language to make the paper easier to read, keeping the most important features, relevant to our main goal, which is, to show how a QB4OLAP cube can be queried without the need of knowing SPARQL programming.

We start the presentation describing the operators, using the Eurostat data cube as our running example.

Operators. The *ROLLUP* operation aggregates measures along a dimension hierarchy to obtain measures at a coarser granularity. The syntax for this operation is:

ROLLUP(CubeName, Dimension, Level)

where Level is the level in Dimension to which the aggregation is performed.

Example 8 (ROLLUP). To compute *the total number of applications by country,* we should write

ROLLUP(Asylum_applications, Citizenship, Country)

The names of the dimensions and levels, are based on the conceptual model in Fig. 1. □

The *DRILLDOWN* operation performs the inverse of ROLLUP; that is, it goes from a more general level to a more detailed level down in a hierarchy. The syntax of this operation is as follows:

DRILLDOWN(CubeName, Dimension, Level)

where Level is the level in Dimension to which the operation is performed.

Example 9 (DRILLDOWN). After rolling-up to the year level, we may want to drill-down to the month level. For that, we write:

DRILLDOWN(YearCube, Time, Month)

Note that we assume that we created the cube YearCube after rolling-up to year. □

The *SLICE* operation removes a dimension from a cube (i.e., a cube of $n-1$ dimensions is obtained from a cube with n dimensions) by selecting one instance in a dimension level. The syntax of this operation is:

SLICE(CubeName, Dimension)

where the Dimension will be dropped by fixing a single value in the Level instance. The other dimensions remain unchanged.

The *DICE* operation returns a cube with the same dimensionality of the original one, but only containing the cells that satisfy a Boolean condition. The syntax for this operation is

DICE(CubeName, Condition)

where Condition is a Boolean condition over dimension levels, attributes, and measures. The *DICE* operation is analogous to a selection in the relational algebra.

Usually, slicing and dicing operations are applied together.

Example 10 (SLICE and DICE). If in our running example we want to remove the Time dimension, we would write:

SLICE(Asylum_applications, Time)

If we want to keep only applications made by Egyptian citizens, we write:

DICE(Asylum_applications,Citizenship.Country.CountryName = 'Egypt')

Note that the dicing condition is applied on the value of a level attribute. This is easier than applying a condition over an IRI, illustrating one of the advantages of supporting level attributes in QB4OLAP. □

We remark that in this paper we limit ourselves to show only the four operations above, since they are enough to illustrate the main idea behind this proposal. A more detailed explanation, and further operations, can be found in [7].

A QL query (or program), is a sequence of OLAP operators, which can store intermediate results in variables bound to cubes, which can be used as arguments in subsequent operations. For example, the following query performs a slicing operation over the Destination dimension, an aggregation to the year level in the Time dimension, and finally filters the result to obtain only the number of asylum applications submitted by citizens from African countries.

```
1  PREFIX data: <http://eurostat.linked-statistics.org/data/>;
2  PREFIX schema: <http://www.fing.edu.uy/inco/cubes/schemas/migr_asyapp#>;
3  QUERY
4  $C1 := SLICE(data:migr_asyappctzm, schema:destinationDim);
5  $C2 := ROLLUP($C1,schema:timeDim,schema:year);
6  $C3 := DICE ($C2, (schema:citizenshipDim|schema:continent|schema:continentName = "Africa"));
```

Note that we have included in the language the Turtle prefixes, which, of course do not belong to the conceptual level, but we think this helps, from a pedagogical point of view, to better convey the idea. In a user-oriented implementation these names can be easily hidden, that is, it would be trivial to write year instead of schema:year.

Finally, to make the presentation simpler, in what follows we assume that QL queries have the following pattern: (ROLLUP | SLICE | DRILLDOWN)* (DICE)*. That means, DICE operators are the last ones in a query, i.e., no ROLLUP, DRILLDOWN or SLICE operations can follow a DICE one.

6.2   Query Simplification

Automatic query simplification and amelioration is important for two reasons: (a) Users will not always write "good" QL queries: although syntactically correct, redundant and/or unnecessary operations could be included; (b) The order in which the operations are written in a query is not always the best one. Thus, a set of rules simplify and ameliorate the queries proposed by users. The **simplification** process deals with the elimination of redundancy in the queries. The **amelioration** process typically aims at producing a query, equivalent to the original one, but which performs better than it. We briefly explain the simplification process next. To organize the discussion we consider two cases:

– Queries that do not contain DICE operators;
– Queries that contain DICE operators.

Queries Not Including a DICE Operation. In this case, we apply the following rules:

– **Rule 1:** Group all the ROLLUP and DRILLDOWN operations over the same dimension, and replace each group of such operations with a single ROLLUP *from the bottom level* of the dimension to the lowest lever indicated in the drill-down operation(s).
– **Rule 2:** If the query contains a SLICE and a sequence of ROLLUP and DRILLDOWN operations over the same dimension, *remove the sequence* of ROLLUPs and DRILLDOWNs and keep only the SLICE.
– **Rule 3:** Reduce intermediate results by performing SLICE operations as soon as possible.

The rationale of the rules is clear. Rule 1 eliminates the ROLLUPs that will be traversed later down in the hierarchy, when performing the DRILLDOWN. Rule 2 addresses the case in which a SLICE removes a dimension that is traversed using ROLLUPs and DRILLDOWNs. In this case, none of the two latter operations will contribute to the result. Rule 3 reduces the size of the intermediate results as early as possible.

Queries Including a DICE Operation. Taking advantage of the assumption that DICE operators are the last ones in a query, we can split the query in two subsets of statements: one that does not contain DICE operators, and another one that is composed only of DICE operators. We can then apply the rules presented above, to the first portion of the query, keeping the statements that involve DICE operators as in the original query.

6.3   QL by Example

In this section we present some examples of QL queries, and their simplification process.

We start the presentation with a query not containing a DICE operation: *Asylum applications by year and continent where the applicant lives*. This is a typical OLAP query, involving two ROLLUP operations, to the Year and Continent levels in dimensions Time and Citizenship.

Query 1: Asylum applications by year and continent.

```
 1  PREFIX schema: <http://www.fing.edu.uy/inco/cubes/schemas/migr_asyapp#>;
 2  PREFIX data: <http://eurostat.linked−statistics.org/data/>;
 3  QUERY
 4  $C1 := ROLLUP (data:migr_asyappctzm, schema:citizenshipDim,schema:continent);
 5  $C2 := ROLLUP ($C1, schema:citizenshipDim,schema:citAll);
 6  $C3 := ROLLUP ($C2, schema:timeDim, schema:year);
 7  $C4 := SLICE ($C3, schema:destinationDim);
 8  $C5 := SLICE ($C4, schema:asylappDim);
 9  $C6 := SLICE ($C5, schema:sexDim);
10  $C7 := SLICE ($C6, schema:ageDim);
11  $C8 := DRILLDOWN ($C7, schema:citizenshipDim,schema:continent);
```

Note that this is not the best way of writing this query, since the ROLLUP to All is clearly not needed (recall that we want to promote the analysis within non-expert OLAP users). Thus, applying Rule 1, the sequence of ROLLUPs and DRILLDOWNs over schema:citizenshipDim dimension is replaced by a single ROLLUP from the bottom level up to the level reached by the last operation in the sequence (in this case schema:continent). The simplified query looks as follows.

```
 1  PREFIX data: <http://eurostat.linked−statistics.org/data/>;
 2  PREFIX schema: <http://www.fing.edu.uy/inco/cubes/schemas/migr_asyapp#>;
 3  QUERY
 4  $C1 := SLICE ($data:migr_asyappctzm, schema:destinationDim);
 5  $C2 := SLICE ($C1, schema:assylappDim);
 6  $C3 := SLICE ($C2, schema:sexDim);
 7  $C4 := SLICE ($C3, schema:ageDim)
 8  $C5 := ROLLUP ($C4, schema:citizenshipDim,schema:continent);
 9  $C6 := ROLLUP ($C5, schema:timeDim, schema:year);
```

Let us now show a query including dicing operations. We want to obtain *Asylum applications by year submitted by Asian citizens, where applications count >5000 whose destination is France or Germany*.

Query 2: Asylum applications by year submitted by Asian citizens, where the number of applications is larger than 5000, and whose destination is France or Germany.

```
 1  PREFIX schema: <http://www.fing.edu.uy/inco/cubes/schemas/migr_asyapp#>;
 2  PREFIX data: <http://eurostat.linked−statistics.org/data/>;
 3  PREFIX property: <http://eurostat.linked−statistics.org/property#>;
 4  PREFIX sdmx−measure: <http://purl.org/linked−data/sdmx/2009/measure#>;
 5  QUERY
 6  $C1 := ROLLUP (data:migr_asyappctzm, schema:citizenshipDim,schema:citAll);
 7  $C2 := ROLLUP ($C1, schema:timeDim, schema:year);
 8  $C3 := DRILLDOWN ($C2, schema:citizenshipDim,schema:continent);
 9  $C4 := SLICE ($C3, schema:asylappDim);
10  $C5 := SLICE ($C4, schema:sexDim);
11  $C6 := SLICE ($C5, schema:ageDim);
```

```
12  $C7 := DICE ($C6, (schema:citizenshipDim|schema:continent|schema:continentName = "Asia"));
13  $C8 := DICE ($C7, ( sdmx−measure:obsValue > 5000 AND
14  (schema:destinationDim|property:geo|schema:countryName = "France") OR
15  (schema:destinationDim|property:geo|schema:countryName = "Germany")));
```

Below, we show the "simplified" query. Again, the sequence of roll-ups and drill-downs is replaced by a roll-up from the bottom level of the hierarchy.

```
1   PREFIX schema: <http://www.fing.edu.uy/inco/cubes/schemas/migr_asyapp#>;
2   PREFIX data: <http://eurostat.linked−statistics.org/data/>;
3   PREFIX property: <http://eurostat.linked−statistics.org/property#>;
4   PREFIX sdmx−measure: <http://purl.org/linked−data/sdmx/2009/measure#>;
5   QUERY
6   $C1 := SLICE (data:migr_asyappctzm, schema:asylappDim);
7   $C2 := SLICE ($C1, schema:sexDim);
8   $C3 := SLICE ($C2, schema:ageDim);
9   $C4 := ROLLUP($C3,schema:timeDim,schema:year);
10  $C5 := ROLLUP($C4,schema:citizenshipDim,schema:continent);
11  $C6 := DICE ($C5(schema:citizenshipDim|schema:continent|schema:continentName = "Asia"));
12  $C7 := DICE ($C6, ( sdmx−measure:obsValue > 5000 AND
13  (schema:destinationDim|property:geo|schema:countryName = "France") OR
14  (schema:destinationDim|property:geo|schema:countryName = "Germany")));
```

7   Translating QL Queries into SPARQL

We expressed above that QB4OLAP provides the metadata needed to automatically translate a high-level language into SPARQL. This is a key feature to promote the use of the semantic web: users would not need to learn a new and complex language like SPARQL. In our case, OLAP users will only need to write relatively simple QL programs, and they will have the flexibility to analyze data cubes on-the-fly.

We now describe a mechanism for translating a QL program into a single SPARQL query. Again, we consider two cases: (1) Queries that do not contain DICE operations, and (2) Queries that contain DICE operations. In Sect. 7.1 we describe the generation of SPARQL queries in the first group, while in Sect. 7.2 we present the rules for the second group of queries.

7.1   Queries Not Including a DICE Operation

After applying the rules presented in the previous section to the original query, we reduce all the possible queries to some kind of "normal form", where, for each dimension D in the data cube only one of the following conditions is satisfied:

– No operation is performed over D
– A ROLLUP operation is performed over D
– A SLICE operation is performed over D

ROLLUPs are implemented navigating the rollup relationships between members, guided by the dimension hierarchy, and aggregations are performed using GROUP BY clauses. The former are performed through SPARQL joins, as we show in the example below. The reader can now better understand why we cannot do this for QB-annotated data sets: they lack the necessary metadata.

SLICEs over dimensions correspond to "slicing out" dimensions. This operation requires measure values to be aggregated up to the ALL level of the dimension being sliced out. The mechanism for this is the same one that is used to compute a ROLLUP.

Therefore, after simplifying and ameliorating the query, we can automatically translate it into a single SPARQL expression.

Example 11. We next show the SPARQL query produced for Query 1.

```
1  PREFIX qb: <http://purl.org/linked−data/cube#>
2  PREFIX qb4o: <http://purl.org/qb4olap/cubes#>
3  PREFIX skos: <http://www.w3.org/2004/02/skos/core#>
4  SELECT ?plm1 ?plm2 (SUM(<http://www.w3.org/2001/XMLSchema#integer>(?m1)) as ?ag1)
5  FROM <http://www.fing.edu.uy/inco/cubes/instances/migr_asyapp_clean>
6  FROM <http://www.fing.edu.uy/inco/cubes/schemas/migr_asyappctzmQB4O>
7  WHERE {
8      ?o a qb:Observation .
9      ?o qb:dataSet <http://eurostat.linked−statistics.org/data/migr_asyappctzm> .
10     ?o <http://purl.org/linked−data/sdmx/2009/measure#obsValue> ?m1 .
11     ?o <http://purl.org/linked−data/sdmx/2009/dimension#refPeriod> ?lm1 .
12     ?lm1 qb4o:memberOf <http://purl.org/linked−data/sdmx/2009/dimension#refPeriod> .
13     ?lm1 <http://www.w3.org/2004/02/skos/core#broader> ?plm1 .
14     ?plm1 qb4o:memberOf <http://www.fing.edu.uy/inco/cubes/schemas/migr_asyapp#year> .
15     ?o <http://eurostat.linked−statistics.org/property#citizen> ?lm2 .
16     ?lm2 qb4o:memberOf <http://eurostat.linked−statistics.org/property#citizen> .
17     ?lm2 <http://www.w3.org/2004/02/skos/core#broader> ?plm2 .
18     ?plm2 qb4o:memberOf <http://www.fing.edu.uy/inco/cubes/schemas/migr_asyapp#continent> .
19  }
20  GROUP BY ?plm1 ?plm2
```

Note that the SLICE operations are implemented omitting, in the SELECT clause, the variables corresponding to the dropped dimensions. Navigation is performed through joins. Lines 8 through 10 (within the WHERE clause), identify the observations, and Line 11 takes the bottom level of the time dimension, which is used to navigate, through the skos:broader predicate, up to the year level. We proceed analogously with the Citizenship dimension: variable ?lm2 is used to navigate the hierarchy up to the continent level, bound to variable ?plm2. Finally, the GROUP BY clause is applied, and an aggregation using function SUM is performed.   □

7.2 Queries Including DICE Operations

In this case, we know that the rules have been applied to the first part of the query, which reduces this part of the query to the cases already described above. The second part of the query contains only DICE operations. Each DICE operation is associated with a condition over measures and/or attribute values, and its result filters out of cells that do not satisfy the condition. We implement the DICE conditions using SPARQL FILTER clauses, also making use of the expressions presented in Sect. 7.1 as subqueries.

The SPARQL query is produced applying the following steps:

1. Obtain a SPARQL query that implements the part of the QL query that does not contain DICE operators, applying the method presented in Sect. 7.1. We will refer to this query as the *inner query*.

2. Produce an *outer SPARQL query* such that:
   (a) Its SELECT clause has the same variables as the SELECT clause of the inner query
   (b) Its WHERE clause contains:
      i. The inner query
      ii. A set of graph patterns to obtain the values of the attributes involved in DICE conditions
      iii. A FILTER clause with the conjunction of the conditions of all the DICE operations

DICE conditions are thus translated into SPARQL expressions. For example, conditions over attributes with range xsd:string are implemented using the REGEX function.

Example 12. This example shows the translation of Query 2, which contains a DICE clause. Here, we use the REGEX clause (which handles regular expressions) within the FILTER condition, to obtain the citizens from Asia, and the destination countries.

```
1  PREFIX qb: <http://purl.org/linked−data/cube#>
2  PREFIX qb4o: <http://purl.org/qb4olap/cubes#>
3  PREFIX skos: <http://www.w3.org/2004/02/skos/core#>
4  SELECT ?ag1 ?plm1 ?lm4 ?plm2
5  WHERE {
6     { ?plm2 <http://www.fing.edu.uy/inco/cubes/schemas/migr_asyapp#continentName> ?plm21 .
7       ?lm4 <http://www.fing.edu.uy/inco/cubes/schemas/migr_asyapp#countryName> ?lm41 .
8     }.
9     { SELECT ?plm1 ?lm4 ?plm2
   (SUM(<http://www.w3.org/2001/XMLSchema#integer>(?m1)) as ?ag1)
10    FROM <http://www.fing.edu.uy/inco/cubes/instances/migr_asyapp_clean>
11    FROM <http://www.fing.edu.uy/inco/cubes/schemas/migr_asyappctzmQB4O>
12    WHERE {
13       ?o a qb:Observation .
14       ?o qb:dataSet <http://eurostat.linked−statistics.org/data/migr_asyappctzm> .
15       ?o <http://purl.org/linked−data/sdmx/2009/measure#obsValue> ?m1 .
16       ?o <http://eurostat.linked−statistics.org/property#age> ?lm1 .
17       ?o <http://purl.org/linked−data/sdmx/2009/dimension#refPeriod> ?lm2 .
18       ?lm2 qb4o:memberOf <http://purl.org/linked−data/sdmx/2009/dimension#refPeriod> .
19       ?lm2 <http://www.w3.org/2004/02/skos/core#broader> ?plm1 .
20       ?plm1 qb4o:memberOf <http://www.fing.edu.uy/inco/cubes/schemas/migr_asyapp#year> .
21       ?o <http://eurostat.linked−statistics.org/property#sex> ?lm3 .
22       ?o <http://eurostat.linked−statistics.org/property#geo> ?lm4 .
23       ?o <http://eurostat.linked−statistics.org/property#citizen> ?lm5 .
24       ?lm5 qb4o:memberOf <http://eurostat.linked−statistics.org/property#citizen> .
25       ?lm5 <http://www.w3.org/2004/02/skos/core#broader> ?plm2 .
26       ?plm2 qb4o:memberOf <http://www.fing.edu.uy/inco/cubes/schemas/migr_asyapp#continent> .
27       ?o <http://eurostat.linked−statistics.org/property#asyl_app> ?lm6 .
28    }
29    GROUP BY ?plm1 ?lm4 ?plm2
30    }
31  FILTER (((REGEX (?plm21,"Asia" , "i")))&&(((?ag1 > 5000) && ((REGEX (?lm41,"France" , "i")) ||
32    (REGEX (?lm41,"Germany" , "i"))))))
33  }
```

Note that the inner and outer queries have the same variables. Also, the outer query contains the FILTER clause, that makes use of the REGEX function. The inner query is solved in the same way as in Example 11.   □

8   Conclusion

In this tutorial paper we have explained how MD data can be represented and queried directly over the SW, without the need of downloading data sets into local DWs. We have shown that, to this end, the RDF Data Cube Vocabulary (QB), the current W3C recommendation must be extended with structural metadata, and dimensional data, in order to be able to support useful OLAP-like analysis. We provided an in-depth comparison between these proposals, and we showed that extending QB with QB4OLAP can be done without re-writing the observations (the largest part of the data). We also presented a high-level query language that allows OLAP users that are not familiar with SW concepts or languages, to write and execute OLAP operators without any knowledge of SPARQL. Queries are automatically translated into SPARQL and executed over an endpoint.

The asylum applications data cube that we have used as running example in this tutorial, as well as an RDF representation of the Northwind DW, and other example cubes, are available at a public SPARQL endpoint.[10] As an exercise, the interested reader can execute the queries presented in this paper, and compare them against the actual Eurostat data, where data are provided in many different ways (reports, graphics, etc.). The analysis allowed by publishing data directly over the SW, using QB4OLAP to represent and enrich data, provides a flexibility that cannot be achieved by traditional publication methods. Moreover, based on the existing observations, expressed in QB, the cost of enriching the original data set is relatively low.

Current work is being carried out along two main lines: (a) Developing further optimization techniques, and providing a benchmark to run queries and study query performance [27,28]; (b) Enhancing usability, by developing semi-automatic techniques to enrich and build existing QB data sets with QB4OLAP metadata [8,9].

Acknowledgments. The author is partially funded by the project PICT 2014 - 0787 CC 0800082612, awarded by the Argentinian Scientific Agency.

References

1. Heath, T., Bizer, C.: Linked Data: Evolving the Web into a Global Data Space. Synthesis Lectures on the Semantic Web. Morgan & Claypool Publishers, San Rafael (2011)
2. Klyne, G., Carroll, J.J., McBride, B.: Resource description framework (RDF): Concepts and abstract syntax (2004). http://www.w3.org/TR/rdf-concepts/
3. Brickley, D., Guha, R., McBride, B.: RDF vocabulary description language 1.0: RDF schema (2004). http://www.w3.org/TR/rdf-schema/

---

[10] https://www.fing.edu.uy/inco/grupos/csi/apps/qb4olap/.

4. Abelló, A., Darmont, J., Etcheverry, L., Golfarelli, M., Mazón, J., Naumann, F., Pedersen, T.B., Rizzi, S., Trujillo, J., Vassiliadis, P., Vossen, G.: Fusion cubes: towards self-service business intelligence. Int. J. Data Warehous. Min. **9**(2), 66–88 (2013)
5. Cyganiak, R., Reynolds, D.: The RDF Data Cube Vocabulary (W3C Recommendation) (2014). http://www.w3.org/TR/vocab-data-cube/
6. Etcheverry, L., Vaisman, A.: QB4OLAP: a vocabulary for OLAP cubes on the semantic web. In: Proceedings of the 3rd International Workshop on Consuming Linked Data, COLD 2012, Boston, USA. CEUR-WS.org (2012)
7. Vaisman, A., Zimányi, E.: Data Warehouse Systems: Design and Implementation. Springer, Heidelberg (2014)
8. Varga, J., Etcheverry, L., Vaisman, A.A., Romero, O., Pedersen, T.B., Thomsen, C.: Enabling OLAP on statistical linked open data. In: Proceedings of the 32nd IEEE International Conference on Data Engineering, ICDE 2016, Helsinki, Finland (2016, to appear)
9. Varga, J., Romero, O., Vaisman, A.A., Etcheverry, L., Pedersen, T.B., Thomsen, C.: Dimensional enrichment of statistical linked open data (2016) (Submitted for publication)
10. Prud'hommeaux, E., Seaborne, A.: SPARQL 1.1 Query Language for RDF (2011). http://www.w3.org/TR/sparql11-query/
11. Etcheverry, L., Vaisman, A.A.: Enhancing OLAP analysis with web cubes. In: Simperl, E., Cimiano, P., Polleres, A., Corcho, O., Presutti, V. (eds.) ESWC 2012. LNCS, vol. 7295, pp. 469–483. Springer, Heidelberg (2012)
12. Etcheverry, L., Vaisman, A., Zimányi, E.: Modeling and querying data warehouses on the semantic web using QB4OLAP. In: Bellatreche, L., Mohania, M.K. (eds.) DaWaK 2014. LNCS, vol. 8646, pp. 45–56. Springer, Heidelberg (2014)
13. Nebot, V., Llavori, R.B.: Building data warehouses with semantic web data. Decis. Support Syst. **52**(4), 853–868 (2012)
14. Kämpgen, B., Harth, A.: Transforming statistical linked data for use in OLAP systems. In: Proceedings of the 7th International Conference on Semantic Systems, I-Semantics 2011, Graz, Austria, pp. 33–40 (2011)
15. Löser, A., Hueske, F., Markl, V.: Situational business intelligence. In: Castellanos, M., Dayal, U., Sellis, T. (eds.) BIRTE 2008. LNBIP, vol. 27, pp. 1–11. Springer, Heidelberg (2009)
16. Ibragimov, D., Hose, K., Pedersen, T.B., Zimányi, E.: Towards exploratory OLAP over linked open data – a case study. In: Castellanos, M., Dayal, U., Pedersen, T.B., Tatbul, N. (eds.) BIRTE 2013 and 2014. LNBIP, vol. 206, pp. 114–132. Springer, Heidelberg (2015)
17. Gómez, L.I., Gómez, S.A., Vaisman, A.A.: A generic data model and query language for spatiotemporal OLAP cube analysis. In: Proceedings of the 15th International Conference on Extending Database Technology, EDBT 2012, pp. 300–311. ACM (2012)
18. Hurtado, C.A., Mendelzon, A.O., Vaisman, A.A.: Maintaining data cubes under dimension updates. In: Proceedings of the 15th International Conference on Data Engineering. ICDE 1999, Sydney, Australia, pp. 346–355. IEEE Computer Society (1999)
19. Vassiliadis, P.: Modeling multidimensional databases, cubes and cube operations. In: Proceedings of the 10th International Conference on Scientific and Statistical Database Management. SSDBM 1998, Capri, Italy, pp. 53–62. IEEE Computer Society (1998)

20. Ciferri, C., Ciferri, R., Gómez, L., Schneider, M., Vaisman, A., Zimányi, E.: Cube algebra: a generic user-centric model and query language for OLAP cubes. Int. J. Data Warehous. Min. **9**(2), 39–65 (2013)
21. SDMX: SDMX standards: Information model (2011). http://sdmx.org/wp-content/uploads/2011/08/SDMX_2-1-1_SECTION_2_InformationModel_201108.pdf
22. Beckett, D., Berners-Lee, T.: Turtle - Terse RDF Triple Language (2011). http://www.w3.org/TeamSubmission/turtle/
23. Hausenblas, M., Ayers, D., Feigenbaum, L., Heath, T., Halb, W., Raimond, Y.: The Statistical Core Vocabulary (SCOVO) (2011). http://vocab.deri.ie/scovo
24. Cyganiak, R., Field, S., Gregory, A., Halb, W., Tennison, J.: Semantic statistics : bringing together SDMX and SCOVO. In: Proceedings of the WWW2010 Workshop on Linked Data on the Web, pp. 2–6. CEUR-WS.org (2010)
25. SDMX: Content Oriented Guidelines (2009). https://sdmx.org/?p=163
26. Bouza, M., Elliot, B., Etcheverry, L., Vaisman, A.A.: Publishing and querying government multidimensional data using QB4OLAP. In: Proceedings of the 9th Latin American Web Congress, LA-WEB 2014, Ouro Preto, Minas Gerais, Brazil, pp. 82–90 (2014)
27. Etcheverry, L., Gómez, S., Vaisman, A.A.: Modeling and querying data cubes on the semantic web (2015). CoRR abs/1512.06080
28. Etcheverry, L., Vaisman, A.A.: Querying semantic web data cubes efficiently (2016) (Submitted for publication)

Design Issues in Social Business Intelligence Projects

Matteo Golfarelli[✉]

DISI, University of Bologna, Bologna, Italy
matteo.golfarelli@unibo.it

Abstract. With the term Social Business Intelligence we refer to a branch of Business Intelligence specialized in applying On-Line Analytical Processing analysis to User-Generated Contents collected from the Web and other sources of social information. The high dynamics of the domain as well as the nature of the source data, that are textual rather than numerical, require specific techniques both for *modeling data* and *managing a project*. Despite the increasing diffusion of Social Business Intelligence applications, only few works in the academic literature addressed such distinguishing features. In this paper we propose both a modeling technique and a methodology that enable the possibility of carrying out a more dynamic and expressive design in Social Business Intelligence projects. We also propose a set of experimental results on real data and real projects proving the effectiveness of our solutions.

1 Introduction

The planetary success of social networks and the widespread diffusion of portable devices has enabled simplified and ubiquitous forms of communication. This in turn has contributed, during the last decade, to a significant shift in human communication patterns towards the *voluntary sharing of personal information*. Most of us are able to connect to the Internet anywhere, anytime, and continuously send messages to a virtual community centered around blogs, forums, social networks, and the like. This has resulted in the accumulation of enormous amounts of *user-generated content* (UGC), that include geolocation, preferences, opinions, news, etc. This huge wealth of information about people's tastes, thoughts, and actions is obviously raising an increasing interest from decision makers because it can give them a fresh and timely perception of the market mood. Besides, the diffusion of UGC is so widespread to directly influence in a decisive way the phenomena of business and society [1–3].

Some commercial tools are available for analyzing the UGC from a few predefined points of view (e.g., topic discovery, brand reputation, and topics correlation) and using some ad hoc *Key Performance Indicators* - KPIs (e.g., topic presence counting and topic sentiment). These tools do not rely on any standard data schema; often they do not even lean on a relational DBMS but rather on in-memory or non-SQL ones. Currently, they are perceived by companies as self-standing applications, so UGC-related analyses are run separately from those

© Springer International Publishing Switzerland 2016
E. Zimányi and A. Abelló (Eds.): eBISS 2015, LNBIP 253, pp. 62–86, 2016.
DOI: 10.1007/978-3-319-39243-1_3

strictly related to business, that are carried out based on corporate data using traditional business intelligence platforms. To give decision makers an unprecedentedly comprehensive picture of the ongoing events and of their motivation, this gap must be bridged [4].

Social Business Intelligence[1] (SBI) is the emerging discipline that aims at effectively and efficiently combining corporate data with UGC to let decisionmakers analyze and improve their business based on the trends and moods perceived from the environment [5]. As in traditional business intelligence, the goal of SBI is to enable powerful and flexible analyses for decision makers with a limited expertise in databases and programming. In other terms we want to apply On-Line Analytical Processing - OLAP analysis on top of a data warehouse storing a semantically enriched version of the UGC related to a specific matter.

In the context of SBI, the most widely used category of UGC is the one coming in the form of textual *clips*. Clips can either be messages posted on social media (such as Twitter, Facebook, blogs, and forums) or articles taken from online newspapers and magazines. Digging information useful for decision makers out of textual UGC requires to set up an *extended* Extraction-Trasformation-Loading - ETL process that includes (1) crawling the Web to extract the clips related to a *subject area*; (2) enriching them in order to let as much information as possible emerge from the raw text; (3) transforming and modeling the data in order to store them in a multidimensional fashion. The subject area defines the project scope and extent, and can be for instance related to a brand or a specific market. Enrichment activities may simply identify the structured parts of a clip, such as its author, or even use *sentiment analysis* techniques [6–8] to interpret each sentence and if possible assign a *sentiment* (also called *polarity*, i.e., positive, negative, or neutral) to it. We call *SBI process* the one that *crawl*, *enrich* and *transform* raw clips in order to let decision makers to carry out in-depth analyses on their content. The SBI process shows several differences with respect to the traditional BI one carried out on *structured* enterprise data (i.e. *owned* data). Such differences, that are discussed in the following sections, impact both on the techniques and methodological steps necessary to transform a raw textual clip in a valuable information.

SBI has emerged as an application and research field in the last few years. Although a wide literature is available on the two initial steps of the extended ETL process sketched so far, namely data crawling, text mining, semantic enrichment and Natural Language Processing, only few papers have focused on the strictly OLAP-related issues. OLAP is based on multidimensional modeling: a *cube* stores a set of measures providing a quantitative evaluation of an event that is defined by a set of dimension of analyses further described at differen level of details by a set of attributes. In [9] the authors propose a cube for analyzing terms occccurrences in documents belonging to a corpus. Due to a the very simple terms categorization approach analysis at different levels of abstraction cannot

---

[1] In the literature the term Social BI is also used to define the collaborative development of post user-generated analytics among business analysts and data mining professionals.

be carried out. In [10] the authors propose textual measures as a solution to summarize textual information within a cube. Complete architectures for SBI have been proposed by [2] and by [11] identifying its basic blocks but still with a limited expressiveness. An important step in increasing the expressiveness of SBI queries has been done in [12] where, a first advanced solution for modeling – the so-called *topic hierarchy* – has been proposed. In this paper we discuss three issues that, in our experience, represent major changes with respect to tradition BI projects:

- SBI Architecture (see Sect. 2): with reference to standard BI projects, SBI requires additional modules necessary, for example, for semantic enrichment of unstructured data. It also requires new technologies such as document DBMS necessary for storing and querying the large amount textual UGC.
- Modeling of SBI data (see Sects. 3 and 4): the semi-structured nature of SBI data together with the dynamism of UGCs make traditional multidimensional models not expressive enough to support SBI queries.
- Methodology for SBI projects (See Sect. 5.a) distinctive feature of SBI projects is related to the huge dynamism of the UGC and of the pressing need of immediately perceiving and timely reacting to changes in the environment.

2   An SBI Architecture

The architecture we propose to support our approach to SBI is depicted in Fig. 1. Its main highlight is the integration between sentiment and business data, which is achieved in a non-invasive way by extracting some business flows from the enterprise data warehouse and integrating them with those carrying textual UGC, in order to provide decision makers with 360° decisional capabilities. In the following we briefly comment each component.

The *Crawling* component carries out a set of keyword-based queries aimed at retrieving the clips (and the available meta-data) that are in the scope of the subject area. The target of the crawler search could be either the whole Web or

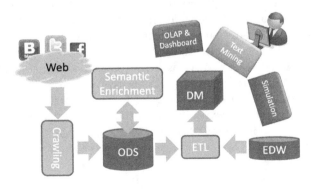

Fig. 1. An architecture for SBI

a set of user-defined web sources (e.g., blogs, forums, web sites, social networks). The semi-structured output of the crawler is turned into a structured form and loaded onto the *Operational Data Store* (ODS), that stores all the relevant data about clips, their authors, and their source channels. To this end, a relational ODS can be coupled with a document-oriented database that can efficiently store and search the text of the clips. The ODS also represents all the topics within the subject area and their relationships. The *Semantic Enrichment* component works on the ODS to extract the semantic information hidden in the clip texts. Depending on the technology adopted (e.g., supervised machine-learning [13] or lexicon-based techniques [14] such information can include the single sentences in the clip, its topic(s), the syntactic and semantic relationships between words, or the sentiment related to a whole sentence or to each single topic it contains. The *ETL* component periodically extracts data about clips and topics from the ODS, integrates them with the business data extracted from the *Enterprise Data Warehouse* (EDW), and loads them onto the *Data Mart* (DM). The DM stores integrated data in the form of a set of multidimensional cubes that, as shown in Sect. 3, require ad hoc modeling solutions; these cubes support the decision making process in three complemental ways:

1. *OLAP & Dashboard*: decision makers can explore the UGC from different perspectives and effectively control the overall social feeling. Using OLAP tools for analyzing UGC in a multidimensional fashion pushes the flexibility of our architecture much further than the standard architectures adopted in this context.
2. *Data Mining*: decision makers evaluate the actual relationship between the rumors/opinion circulating on the Web and the business events (e.g., to what extent positive opinions circulating about a product will have a positive impact on sales?).
3. *Simulation*: the correlation patterns that connect the UGC with the business events, extracted from past data, are used to forecast business events in the near future given the current UGC.

In our prototypical implementation of this architecture, publicly available at http://semantic.csr.unibo.it, topics and roll-up relationships are manually defined; we use Brandwatch for keyword-based crawling, Talend for ETL, SyN Semantic Center by SyNTHEMA for semantic enrichment (specifically, for labeling each clip with its sentiment), Oracle for storing the ODS and the DM, and MongoDB for storing the document database. We developed an ad hoc OLAP & dashboard interface using JavaScript, while simulation and data mining components are not currently implemented.

The architectural components mentioned above are normally present, though with different levels of sophistication, in most current commercial solutions for SBI. However the roles in charge of designing, tuning, and maintaining each component may vary from project to project. In regards to this, SBI projects can be classified as follows:

- *Level 1: Best-of-Breed.* In this type of projects, a best-of-breed policy is followed to acquire tools specialized in one of the steps necessary to transform raw clips in semantically-rich information. This approach is often followed by those who run a medium to long-term project to get full control of the SBI process by finely tuning all its critical parameters, typically aimed at implementing ad hoc reports and dashboards to enable sophisticated analyses of the UGC.
- *Level 2: End-to-End.* Here, an end-to-end software/service is acquired and tuned. Customers only need to carry out a limited set of tuning activities that are typically related to the subject area, while a service provider or a system integrator ensures the effectiveness of the technical (and domain-independent) phases of the SBI process.
- *Level 3: Off-the-Shelf.* This type of projects consists in adopting, typically in a *as-a-service* manner, an off-the-shelf solution supporting a set of reports and dashboards that can satisfy the most frequent decision makers needs in the SBI area (e.g., average sentiment, top topics, trending topics, and their breakdown by source/author/sex). With this approach the customer has a very limited view of the single activities that constitute the SBI process, so she has little or no chance of positively impacting on activities that are not directly related to the analysis of the final results.

Moving from level 1 to 3, projects require less technical capabilities from customers and ensure a shorter set-up time, but they also allow less control of the overall effectiveness and less flexibility in analyzing the results.

3   Modeling SBI Data

The main goal of SBI is to allow OLAP paradigm to be applied to social/textual data. As shown in the previous section some proposals for a multidimensional modeling of SBI data have been provided but all of them lack in providing the required expressiveness. A key role in the analysis of textual UGC is played by *topics*, meant as specific concepts of interest within the subject area. Decision makers are interested in knowing how much people talk about a topic, which words are related to it, if it has a good or bad reputation, etc. Thus, topics are obvious candidates to become a dimension of the cubes for SBI. A simple example of an SBI cube is reported in Fig. 2. Apart from the Topic hierarchy, the meta-data retrieved by the crawling module has been modeled thus, for example, the average sentiment about a specific group of topics can be analyzed for different Media Types (e.g. Forum, News). Like for any other dimension, decision makers are very interested in grouping topics together in different ways to carry out more general and effective analyses—which requires the definition of a topic hierarchy that specifies inter-topic roll-up (i.e., grouping) relationships that, in turns, enable aggregations of topics at different levels.

Example 1. A marketing analyst wants to analyze people's feelings about mobile devices and relate them to the selling trends. A basic cube she would use to this

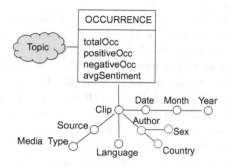

Fig. 2. A Dimensional Fact Model - DFM representation of an SBI cube reporting information about the occurrence of a specific topic in a specific text.

purpose is the one counting, within the textual UGC, the number of occurrences of each topic related to subject area "mobile technologies", distinguishing between those expressing positive/negative sentiment as labeled by an opinion mining algorithm (see Fig. 2). Figure 3-right shows a set of topics for mobile technologies and their roll-up relationships: when computing the brand reputation for the topic "Samsung", decision makers may wish to also include occurrences of topics "Galaxy III" and "Galaxy Tab", while when analyzing decision makers' concerns about "Galaxy III" she wants to consider comments about its parts.

However, topic hierarchies are different from traditional hierarchies (like the temporal and the geographical one) in several ways:

1. Non-leaf topics can be related to facts too(e.g., clips may talk of smartphones as well as of the Galaxy III) [12]. This means that grouping topics at a given level may not determine a total partitioning of facts [15]. Besides, topic hierarchies are unbalanced, i.e., hierarchy instances can have different lengths.
2. Trendy topics are heterogeneous (e.g., they could include names of famous people, products, places, brands, etc.) and change quickly over time (e.g., if at some time it is announced that using smartphones can cause finger pathologies, a brand new set of hot unpredicted topics could emerge during the following days), so a comprehensive schema for topics cannot be anticipated at design time and must be dynamically defined.
3. Roll-up relationships between topics can have different semantics: for instance, the relationship semantics in "Galaxy III has brand Samsung" and "Galaxy III has type smartphone" is quite different. In traditional hierarchies this is indirectly modeled by leaning on the semantics of aggregation levels ("Smartphone" is a member of level Type, "Samsung" is a member of level Brand).

In light of the above, topic hierarchies in Relational OLAP (ROLAP) contexts must clearly be modeled with more sophisticated solutions than traditional star schemata. In [16] we proposed *meta-star*; its basic idea is to use meta-modeling coupled with navigation tables and with traditional dimension tables. On the one hand, navigation tables easily support hierarchy instances with different lengths and with non-leaf facts (requirement ♯1), and allow different roll-up semantics to be explicitly annotated (requirement ♯3); on the other, meta-modeling enables hierarchy heterogeneity and dynamics to be accommodated (requirement ♯2). An obvious consequence of the adoption of navigation tables is that the total size of the solution increases exponentially with the size of the topic hierarchy. This clearly limits the applicability of the meta-star approach to topic hierarchies of small-medium size; however, we argue that this limitation is not really penalizing because topic hierarchies are normally created and maintained manually by domain experts, which suggests that their size can hardly become too large.

In the remainder of this section we provide a formal definition of the topic hierarchy related concepts.

Definition 1. *A hierarchy schema* S *is a couple of a set* L *of levels and a roll-up partial order* $\succ$ *of* L. *We write* $l_k \dot{\succ} l_j$ *to emphasize that* l_k *is an immediate predecessor of* l_j *in* $\succ$.

Example 2. In Example 1 it is $L = \{$Product, Type, Category, Brand, Component$\}$ and Component $\dot{\succ}$ Product $\dot{\succ}$ Type $\dot{\succ}$ Category, Product $\dot{\succ}$ Brand (see Fig. 3-left).

The connection between hierarchy schemata (intension) and topic hierarchies (extension) is captured by Definition 2, that also annotates roll-up relationships with their semantics.

Definition 2. *A topic hierarchy conformed to hierarchy schema* $S = (L, \succ_S)$ *is a triple of (i) an acyclic directed graph* $H = (T, R)$, *where* T *is a set of topics and* R *is a set of inter-topic roll-up relationships; (ii) a partial function* $Lev : T \rightarrow L$ *that associates some topics to levels of* S; *and (iii) a partial function* $Sem : R \rightarrow \rho$ *that associates some roll-up relationships to their semantics (with* ρ *being a list of user-defined roll-up semantics). Graph* H *must be such that, for each ordered pair of topics* $(t_1, t_2) \in R$ *such that* $Lev(t_1) = l_1$ *and* $Lev(t_2) = l_2$, *it is* $l_1 \dot{\succ} l_2$ *and* $\forall (t_1, t_3) \in R, Lev(t_3) \neq l_2$.

Example 3. In Fig. 3 the topic hierarchy on the right-hand side is annotated with levels and roll-up semantics; T includes for instance $\{$*8MP Camera, Galaxy III, Samsung*$\}$; R includes for instance, $\{($*8MP Camera, Galaxy III*$), ($*Galaxy III, Samsung*$)\}$; Lev includes for instance $\{($*8MP Camera,* Component$), ($*Galaxy III,* Product$), ($*Samsung,* Brand$)\}$; finally, Sem includes for instance $(($*8MP Camera, Galaxy III*$),$ *isPartOf*$)$;

The intuition behind the constraints on H in Definition 2 is that inter-topic relationships must not contradict the roll-up partial order and must have many-to-one multiplicity. For instance in Fig. 3, arc from "Galaxy III" to "Smartphone" is correct because Product $\dot{\succ}$ Type, but there could be no other arc from "Galaxy III" to a topic of level Type. In the same way, no arc from a product to a category is allowed; the arc from "Galaxy III" to "Touchscreen" is allowed because the latter does not belong to any level.

Finally, Definition 3 provides a compact representation for the semantics involved in any path of a topic hierarchy.

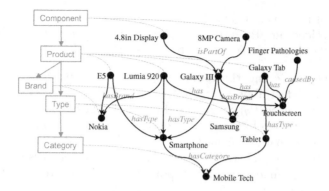

Fig. 3. The annotated topic hierarchy for the mobile technology subject area.

Definition 3. *Given topic t_1 such that $Lev(t_1) = l_1$ and given level l_2 such that $l_1 \succ l_2$, we denote with $Anc^{l_2}(t_1)$ the topic t_2 such that $Lev(t_2) = l_2$ and t_2 is reached from t_1 through a directed path P in H. The roll-up signature of couple (t_1, t_2) is a binary string of $|\rho|$ bits, where each bit corresponds to one roll-up semantics and is set to 1 if at least one roll-up relationship with that semantics is part of P, is set to 0 otherwise. Conventionally, the roll-up signature of (t, t) is a string of 0's for each t.*

Example 4. With reference to Fig. 3 it is for instance $Anc^{\mathsf{Brand}}(8MP\ Camera) = Samsung$, $Anc^{\mathsf{Type}}(8MP\ Camera) = Smartphone$. Note that topics "Touchscreen" and "Finger Pathologies" do not belong to any level. If $\rho = (isPartOf, hasType, hasBrand, hasCategory, has, causedBy)$, then the roll-up signature of *(8MP Camera, Samsung)* is 101000 (because the path from "8MP Camera" to "Samsung" includes roll-up relationships with semantics *isPartOf* and *hasBrand*), that of *(8MP Camera, Smartphone)* is 110000.

Topic hierarchies can be implemented on a ROLAP platform combining classical dimension tables with recursive navigation tables and extends the result by meta-modeling. Remarkably, the designer can tune the solution by deciding which levels $L^{stat} \subseteq L$ are to be modeled also in a static way, i.e., like in a classical dimension table. Two different tables are used:

1. A *topic table* storing one row for each distinct topic $t \in T$. The schema of this table includes a primary surrogate key IdT, a Topic column, a Level column, and an additional column for each static level $l \in L^{stat}$. The row associated to topic t has Topic$= t$ and Level$= Lev(t)$. Then, if $Lev(t) \in L^{stat}$, that row has value t in column $Lev(t)$, value $Anc^l(t)$ in each column l such that $l \in L^{stat}$ and $Lev(t) \succ l$, and NULL elsewhere.
2. A *roll-up table* storing one row for each topic in T and one for each arc in the transitive closure of H. The row corresponding to topic t has two foreign keys, ChildId and ParentId, that reference the topic table and both store the surrogate of topic t, and a column RollUpSignature that stores the roll-up

signature of (t, t), i.e., a string of 0's. The row corresponding to arc (t_1, t_2) stores in ChildId and ParentId the two surrogates of topics t_1 and t_2, while column RollUpSignature stores the roll-up signature of (t_1, t_2).

Example 5. The topic and the roll-up tables for the topic hierarchy in Fig. 3 when $L^{stat} = \{$Product, Type, Category$\}$ and $\rho = (isPartOf, hasType, hasBrand, hasCategory, has, causedBy)$ are reported in Fig. 4. The eleventh row of the roll-up table states that the roll-up signature of couple *(8MP Camera, Smartphone)* is 110000, i.e., that the path from one topic to the other includes semantics *isPartOf* and *hasType*. To achieve a better understanding of the differences between meta-star and star schema modeling, Fig. 5 reports the two complete logical schemata.

Meta-stars also better support topic hierarchy dynamics, through the combined use of meta-modeling and of the roll-up table. A whole new set of emerging topics, possibly structured in a hierarchy with different levels, can be accommodated—without changing the schema of meta-stars—by adding new values to the domain of the Level column, adding rows to the topic and the roll-up tables to represent the new topics and their relationships, and extending the

TOPIC_T

IdT	Topic	Level	Product	Type	Category
1	8MPCamera	Component	–	–	–
2	GalaxyIII	Product	GalaxyIII	Smartph.	MobTech
3	GalaxyTab	Product	GalaxyTab	Tablet	MobTech
4	Smartphone	Type	–	Smartph.	MobTech
5	Tablet	Type	–	Tablet	MobTech
6	MobileTech	Category	–	–	MobTech
7	Samsung	Brand	–	–	–
8	Finger Path.	–	–	–	–
9	Touchscreen	–	–	–	–
...	...	...	...	...	...

ROLLUP_T

ChildId	ParentId	RollUpSignature
1	1	000000
2	2	000000
...	...	000000
1	2	100000
2	4	010000
2	7	001000
4	6	000100
8	9	000001
2	9	000010
...	...	...
1	4	110000
1	7	101000
1	9	100010
2	6	010100
3	6	010100
...	...	...
1	6	110100
...	...	...

Fig. 4. Meta-star modeling for the mobile technology subject area depicted in Fig. 3, using $\rho = (isPartOf, hasType, hasBrand, hasCategory, has, causedBy)$.

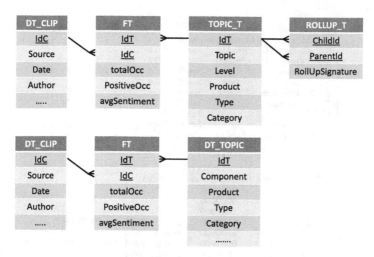

Fig. 5. Meta-star logical model (top) and standard star schema model (botton) for the running example.

roll-up signatures with new bits for the new roll-up semantics. The newly-added levels will immediately become available for querying and aggregation.

3.1   Slowly-Changing Topics and Levels

Historicization of hierarchies is a relevant issue in data warehouse design since it allows decision makers to better focus their analyses and enables the execution of queries that use different hierarchy versions. Hierarchies subject to changes in their data are normally referred to as *slowly-changing dimensions* [17], and different techniques can be adopted to cope with them. In particular, in a star schema implementation of a cube, a *Type-2* solution is one where data versions are tracked by creating multiple tuples in the dimension table for the same natural key (e.g., several tuples in the product dimension table corresponding to different classifications into types of the same product at different times); each fact in the fact table is then referred to the specific tuple that was valid when the fact took place, so that the historical truth can easily be reconstructed by a simple star join. A more powerful solution is so-called *full logging* [18], that adds a couple of time stamps to dimension tables to explicitly model the temporal validity of each version so as to enable more expressive queries. While handling different data versions is essentially a technical problem, dealing with changes in the *schema* of hierarchies is still a research issue, with only a few proposed solutions in the literature (e.g., [19]).

Although meta-stars natively support data and schema changes, keeping track of the different versions requires some further expedient. First of all we observe that, thanks to meta-modeling and differently from traditional star schemata, meta-stars can track also schema changes using the same solutions devised for slowly-changing dimensions. This means that not only data changes

TOPIC_T

IdT	Topic	Level	From	To	Master
1	8MPCamera	Component	Jan 01 2014	-	1
2	GalaxyIII	Product	Jan 01 2014	-	2
3	GalaxyTab	Product	Jan 01 2014	-	3
4	Smartphone	Type	Jan 01 2014	Jan 31 2014	4
5	Tablet	Type	Jan 01 2014	-	5
6	MobileTech	Category	Jan 01 2014	-	6
7	Samsung	Brand	Jan 01 2014	-	7
8	Finger Path.	–	Jan 01 2014	-	8
9	Touchscreen	–	Jan 01 2014	-	9
10	Smartphone	SubCat	Feb 1 2014	-	4
...	...	... ...	...	...	...

ROLLUP_T

ChildId	ParentId	RollUpSignature	From	To
1	1	0000000	Jan 01 2014	-
...	...	0000000	...	...
1	2	1000000	Jan 01 2014	-
2	4	0100000	Jan 01 2014	Jan 31 2014
2	10	0100000	Feb 01 2014	-
4	6	0001000	Jan 01 2014	Jan 31 2014
10	6	0001000	Feb 01 2014	-
...	...	...	...	...
1	4	1100000	Jan 01 2014	Jan 31 2014
1	10	1100000	Feb 01 2014	-
...	...	...	...	...

Fig. 6. Full-logging solution for the mobile technology subject area

(i.e., creation of a new member, deletion of a member, and inter-member relationship update), but even schema changes (i.e., creation of a new level, deletion of a level)[2] can be tracked in a meta-star without affecting the schema of topic and roll-up tables.

Both Type-2 and full-logging solutions can be applied to meta-stars. As in star schemata, a Type-2 solution does not impact on the meta-star schema and is implemented by properly setting the ETL process only. Conversely, full logging impacts on the meta-star schema; more precisely, tracking changes in the roll-up partial order requires timestamps in the roll-up table only, while all the other operations also involve the topic table since a change in a topic/level must be reflected in all the related arcs of H^+.

Example 6. A full-logging solution for our motivating example is shown in Fig. 6. On Jan. 31, 2014 a change in the hierarchy schema occurred: level SubCategory was introduced and topic "Smartphone" was moved from Type to SubCategory. A new tuple with IdT 10 was added to TOPIC_T, while the previous version (the tuple with IdT 4) run out of validity; the related arcs in ROLLUP_T were updated accordingly.

---

[2] Though the roll-up partial order between levels is part of the hierarchy schema, its historicization is handled at the instance level in both stars and meta-stars; while from the extensional point of view inter-level relationships can be reconstructed from the relationships between level members, from the intensional point of view they are explicitly stored only in meta-data repositories, not in dimension table schemata.

Though Type-2 and full-logging solutions are more powerful when applied to meta-stars, they should be used carefully because their impact on cardinality of roll-up tables is very strong. Indeed, a roll-up table explicitly stores the transitive closure of inter-topic relationships, so any change in a topic, a level, or an arc may affect several tuples.

4 Querying Meta-Stars

A classical OLAP query includes a group-by clause and a selection clause. In this section we show how meta-stars support OLAP queries with increasing expressiveness and complexity, starting from queries using only static levels to end-up with semantics-aware queries. We preliminarily recall that, in this context, facts can also be associated to non-leaf topics. As a consequence, multiple semantics of aggregation are made available to decision makers. For instance, computing the number of occurrences of "Smartphone" may either mean considering only the UGC mentioning the word "Smartphone", or also considering the UGC mentioning products of type smartphones (such as Galaxy III), or also considering the UGC mentioning a component of a product of type smartphone (such as 8MP Camera).

The queries discussed below are based on the relational implementation of the OCCURRENCE cube (see Fig. 2). The relational schema includes, besides the tables discussed so far (i.e. TOPIC_T, ROLLUP_T), tables DTCLIP and FT. The first one is a separate dimension table storing clips; the second one is the fact table including the occurrence metrics (e.g. totalOcc, avgSentiment).

4.1 Queries without Topic Aggregation

In this family of queries the topic hierarchy is not navigated, i.e., only occurrences of the very topics of interest are counted. These queries can be always formulated on the topic table by relying on the Level column; for instance, the number of total occurrences for each brand on a given date are obtained as follows:

```
SELECT     TOPIC_T.Topic, SUM(FT.TotalOcc)
FROM       TOPIC_T, DTCLIP, FT
WHERE      FT.IdT = TOPIC_T.IdT AND FT.IdC = DTCLIP.IdC AND
           TOPIC_T.Level ="Brand" AND DTCLIP.Date = "06/22/2013"
GROUP BY   TOPIC_T.Topic;
```

Clearly, if the required topic level has been modeled as static, like Type, the query can also be equivalently formulated by directly including that level in the group-by clause:

```
SELECT     TOPIC_T.Type, SUM(FT.TotalOcc)
FROM       TOPIC_T, DTCLIP, FT
WHERE      FT.IdT = TOPIC_T.IdT AND FT.IdC = DTCLIP.IdC AND
           TOPIC_T.Level = "Type" AND DTCLIP.Date = "06/22/2013"
GROUP BY   TOPIC_T.Type;
```

4.2   Queries with Topic Aggregation

In this family of queries the topic hierarchy is extensively navigated, i.e., each topic of interest is considered together with its descendants when computing the number of occurrences. The portion of topic hierarchy that has been modeled as static is easily navigated using the topic table as if it were a classical dimension table; for instance,

```
SELECT  SUM(FT.TotalOcc)
FROM    TOPIC_T, DTCLIP, FT
WHERE   FT.IdT = TOPIC_T.IdT AND FT.IdC = DTCLIP.IdC AND
        TOPIC_T.Category = "Mobile Tech" AND
        DTCLIP.Date = "06/22/2013";
```

returns the occurrences of "Mobile Tech" counting its types and products (but not its components, because $\mathsf{Component} \notin L^{stat}$.

On the other hand, if aggregation is to involve levels that have not been modeled as static, the roll-up table must be used. For instance, this is the case for the talking volume analysis of Example 1, that returns the total number of occurrences for "Mobile Tech" and all its descendants also including components:

```
SELECT  SUM(FT.totalOcc)
FROM    TOPIC_T, ROLLUP_T, DTCLIP, FT
WHERE   FT.IdT = ROLLUP_T.ChildId AND
        ROLLUP_T.ParentId = TOPIC_T.IdT AND
        FT.IdC = DTCLIP.IdC AND
        TOPIC_T.Topic ="Mobile Tech" AND
        DTCLIP.Date = "06/22/2013";
```

In case the desired aggregation includes two or more levels of the topic hierarchy, aliases must be introduced to use different "versions" of the topic and roll-up tables. For instance, the query below computes the average sentiment for each combination of brand and type:

```
SELECT    T1.Topic AS Brand, T2.Topic AS Type, AVG(FT.avgSentiment)
FROM      TOPIC_T T1, ROLLUP_T R1,
          TOPIC_T T2, ROLLUP_T R2, FT
WHERE     FT.IdT = R1.ChildId AND R1.ParentId = T1.IdT AND
          FT.IdT = R2.ChildId AND R2.ParentId = T2.IdT AND
          T1.Level ="Brand" AND T2.Level = "Type"
GROUP BY  T1.Topic, T2.Topic;
```

4.3   Queries with Semantics-Aware Topic Aggregation

While the two previous types of queries can also be formulated on a classical star schema extended with a navigation table to model recursion, this type of query uses the user-defined roll-up semantics to filter the way the topic hierarchy is navigated so as to produce custom aggregations. For instance, this is the case with the brand reputation analysis of Example 1, that returns the number of positive and negative occurrences of each brand and of its products:

```
SELECT    TOPIC_T.Topic, SUM(FT.positiveOcc), SUM(FT.negativeOcc)
FROM      TOPIC_T, ROLLUP_T, FT
WHERE     FT.IdT = ROLLUP_T.ChildId AND
          ROLLUP_T.ParentId = TOPIC_T.IdT AND
          TOPIC_T.Level = "Brand" AND
          ROLLUP_T.RollUpSignature = 001000
GROUP BY  TOPIC_T.Topic;
```

Another query of this family is the one for health rumors analysis, that returns the negative occurrences for touchscreens and the related pathologies:

```
SELECT  TOPIC_T.Topic, SUM(FT.negativeOcc)
FROM    TOPIC_T, ROLLUP_T, FT
WHERE   FT.IdT = ROLLUP_T.ChildId AND
        ROLLUP_T.ParentId = TOPIC_T.IdT AND
        TOPIC_T.Topic = "Touchscreen" AND
        ROLLUP_T.RollUpSignature = 000001;
```

4.4 Evaluation

In this section we evaluate the performance of meta-stars by comparing the efficiency of query execution against star schemata. All tests were conducted using the Oracle 11g RDBMS on a 64-bits AMD Opteron quad-core 2.09 GHz virtual machine, with 4 GB RAM, running Windows Server 2008 R2 Standard SP1.

To conduct the tests we generated a benchmark of sample cubes with different characteristics but all conformed to the conceptual schema of Fig. 2. We created three perfectly height-balanced topic hierarchies with $L^{stat} \equiv L$, in order to create equivalent structures for both the meta-star and the star schema. The parameters used to create the topic hierarchies are the number of levels and the fan-out of each node (i.e., the number of children connected to each father). Table 1 summarizes the characteristics of the topic hierarchies; clearly, the number of topics and the size of the roll-up table increase exponentially with the tree height. In addition, we generated two fact tables, $FT1$ and $FT2$, with 1M and 10M facts respectively, and linked each of them to the previously defined topic tables. For a realistic and fair evaluation, we created B$^+$-indexes on all foreign keys, on the Level column, and on all columns corresponding to static levels; no materialized views were created.

To define the workload for evaluation we considered the query family described in Sect. 4.2 (i.e., the ones based on topic aggregation), that are equally executable on both meta-stars and star schemata and represent the worst case for meta-stars efficiency since they require access to the roll-up table. In particular, we created queries with an increasing number of levels (from 0 to 2) in the group-by clause, in order to evaluate the cost of using one or more roll-up table aliases. The query execution results are shown in Table 2; each execution time displayed is the average time required to run three different queries with the same number of levels in their group-by's and different selection predicates.

Table 1. Characteristics of meta-stars

Topic hier.	TOPIC_T	ROLLUP_T	fan-out	tree height
$H1$	106	626	4	4
$H2$	658	4,514	8	4
$H3$	27,306	334,962	4	8

Table 2. Execution time of queries (in seconds)

Table	Group-by	FT1		FT2	
		Meta-star	Star s.	Meta-star	Star s.
H1	0	13.8	12.7	140.0	137.2
	1	16.0	5.8	174.6	64.3
	2	16.6	14.6	162.4	162.1
H2	0	13.6	13.0	136.0	133.6
	1	16.7	5.6	179.5	179.4
	2	17.0	16.2	175.8	162.2
H3	0	12.2	9.0	139.1	126.6
	1	15.9	14.1	147.3	172.1
	2	35.1	16.9	187.1	144.2

It could be expected that the toll to be paid for increasing querying expressiveness and schema dynamicity, is in terms of performances. Though, as expected, in most cases star schemata outperform meta-stars, the time execution gap is quite limited and perfectly acceptable in terms of *on-line* querying. The gap is significantly smaller, in relative terms, for *FT*2 since the execution time is mostly spent to access the fact table rather than the topic hierarchy. Noticeably, execution times for meta-stars increase smoothly for group-by's with increasing number of levels. The execution time behaves similarly when the cardinality of the topic and roll-up tables increases. In particular, an in-depth analysis of the Oracle execution plans has shown that, although the roll-up table cardinality increases exponentially with the depth of the topic hierarchy (see Table 1), the execution time increases smoothly because indexes allow only the relevant part of that table to be accessed when querying.

The experimental evidences (see [5,16] for more tests) clearly show that meta-star are a valuable technique for increasing the OLAP cube expressiveness while enabling at the same time a more dynamic and agile design. Although, meta-star has been specifically designed for the SBI context, it is a general-purpose solution suitable for many other contexts. In particular, they can be useful to reduce the maintenance costs in all the projects were it is not possible to know at design time the set of dimensional levels.

5   A Methodology for SBI Projects

The availability of more expressive and dynamic modeling techniques is useless without the adoption of a proper design methodology that makes it possible to properly and timely collect new requirements. SBI has emerged as an application and research field in the last few years and there is no agreement yet on how to organize the different design activities. Indeed, in real SBI projects, practitioners typically carry out a wide set of task but they lack an organic and structured view of the design process. The specificities that distinguish a BI project from an SBI one are listed below:

- SBI projects call for an effective and efficient support to maintenance iterations, because of the huge dynamism of the UGC and of the pressing need of immediately perceiving and timely reacting to changes in the environment.
- The schema of the data and the ETL flows are independent of the project domain and the changes are mainly related to the meta-data made available by the crawling and the semantic enrichment engines.
- The complexity of different tasks and the subjects who are in charge of them are strongly related to the type of project implemented.

The iterative methodology we have proposed in [20] (see Fig. 7) is aimed at letting harmoniously coexist all the activities involved in an SBI project. These activities are to be carried out in tight connection one to each other, always keeping in mind that each of them heavily affects the overall system performance and that a single problem can easily neutralize all other optimization efforts.

Besides speeding up the initial design of an SBI process, the methodology is aimed at maximizing the effectiveness of the decision maker analyses by continuously optimizing and refining all its phases. These maintenance activities are necessary in SBI projects because of the continuous environment variability which asks for high responsiveness. This variability impacts every single activity, from crawling design to semantic enrichment design, and leads to constantly having to cope with changes in requirements.

In the following we briefly describe the main feature of each activity, for a more detailed description refer to [20].

1. *Macro-Analysis*: during this activity, decision makers are interviewed to define the project scope and the set of inquiries the system will answer to. An *inquiry* captures an informative need of a decision maker; from a conceptual point of view it is specified by three components: *what*, i.e., one or more topics on which the inquiry is focused (e.g., the Galaxi III); *how*, i.e., the type of analysis the decision maker is interested in (e.g., top related topics); *where*, i.e., the data sources to be searched (e.g., the Technology-related web forums).

   Inquiries drive the definition of subject area, themes, and topics. As said before, the *subject area* of a project is the domain of interest for the decision maker (e.g., Mobile Technology), meant as the set of themes about which information is to be collected. A *theme* (e.g., Tablet reputation) includes a set of specific *topics* (e.g., Touchscreen). Laying down themes and topics at this early stage is useful as a foundation for designing a core taxonomy of topics during the first iteration of ontology design; themes can also be used to enforce an incremental decomposition of the project. In practice, this activity should also produce a first assessment of which sources cannot be excluded from the source selection activity since they are considered as extremely relevant (e.g., the corporate website and Facebook pages).

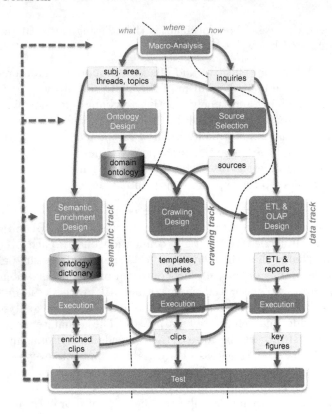

Fig. 7. Functional view of our methodology for SBI design

2. *Ontology Design*: during this activity, customers work on themes and topics to build and refine the domain ontology that models the subject area. Noticeably, the domain ontology is not just a list of keywords; indeed, it can also model relationships (e.g., hasKind, isMemberOf) between topics. Once designed, this ontology becomes a key input for almost all process phases: semantic enrichment relies on the domain ontology to better understand UGC meaning; crawling design benefits from topics in the ontology to develop better crawling queries and establish the content relevance; ETL and OLAP design heavily uses the ontology to develop more expressive, comprehensive, and intuitive dashboards.

3. *Source Selection*: is aimed at identifying as many web domains as possible for crawling. The set of potentially relevant sources can be split in two families: *primary sources* and *minor sources*. The first set includes all the sources mentioned during the first macro-analysis iteration, namely: (1) the corporate communication channels (e.g. the corporate website, Facebook page, Twitter account); (2) the *generalist* sources, such as the online version of the major publications. The user-base of minor sources is smaller but not less relevant to the project scope. Minor sources include lots of small platforms which

produce valuable information with high informative value because of their major focus on themes related to the subject area. The two main subsequent tasks involved in this activity are:

- *Template design* consists in an analysis of the code structure of the source website to enable the crawler to detect and extract only the informative UGC (e.g., by excluding external links, advertising, multimedia, and so on).
- Based on the templates designed, *query design* develops a set of crawling queries to extract the relevant clips. Normally, these are complex keyword-based queries that explicitly mention both relevant keywords to extract on-topic clips and irrelevant keywords to exclude off-topic clips.

Note that filtering off-topic clips at crawling time could be difficult due to the limitations of the crawling language, and also risky because the in-topic perimeter could change during the analysis process. For these reasons, the team can choose to release some constraints aimed at letting a wider set of clips "slip through the net", and only filter them at a later stage using the search features of the underlying document DBMS (e.g., MongoDB).

4. *Semantic Enrichment Design*: involves several tasks whose purpose is to increase the accuracy of text analytics so as to maximize the process effectiveness in terms of extracted *entities* and sentiment assigned to clips; entities are concepts that emerge from semantic enrichment but are not part of the domain ontology yet (for instance, they could be emerging topics). The specific tasks to be performed depend on the semantic engine adopted and on how semantic enrichment is carried out.

In general, two main tasks that enrich and improve its linguistic resources can be distinguished:

- *Dictionary enrichment*, that requires including new entities missing from the dictionary and changing the sentiment of entities (*polarization*) according to the specific subject area (e.g., in "I always eat fried cutlet", the word "fried" has a positive sentiment, but in the food market area a sentence like "These cutlets taste like fried" should be tagged with a negative sentiment because fried food is not considered to be healthy).
- *Inter-word relation definition*, that establishes or modifies the existing semantic, and sometimes also syntactic, relations between words. Relations are linguistically relevant because they can deeply modify the meaning of a word or even the sentiment of an entire sentence determining the difference between right and wrong interpretation (e.g., "a Pyrrhic victory" has negative sentiment though "victory" is positive).

Modifications in the linguistic resources may produce undesired side effects; so, after completing these tasks, a *correctness analysis* should be executed aimed at measuring the actual improvements introduced and the overall ability of the process in understanding a text and assigning the right sentiment to it. This is normally done, using regressive test techniques, by manually tagging an incrementally-built sample set of clips with a sentiment.

5. *ETL & OLAP Design*: the main tasks in this activity are:
   - *ETL design and implementation*, that strongly depends on features of the semantic engine, on the richness of the meta-data retrieved by the crawler

(e.g., URLs, author, source type), and on the possible presence of specific data acquisition channels such as CRM.

- *KPI design*; different kinds of KPIs can be designed and calculated depending on which kinds of meta-data the crawler fetches.
- *Dashboard design*, during which a set of reports is built that captures the decision maker needs expressed by inquiries during macro-analysis.

6. *Execution and Test*: have a basic role in the methodology, as it triggers a new iteration in the design process. Crawling queries are executed, the resulting clips are processed, and the reports are launched over the enriched clips. The specific tests related to each single activity, described in the preceding subsections, can be executed separately though they are obviously inter-related. The first test executed is normally the one of crawling; even after a first round, the semantic enrichment tests can be run on the resulting clips. Similarly, when the first enriched clips are available, the test of ETL and OLAP can be triggered.

The analysis of the outcomes of a set of case studies [20] has shown that the adoption of a proper methodology strongly impacts on the capability of keeping under control execution time, required resources and effectiveness of the results. In particular the key points of the proposed methodology are: (1) a clear organization of goals and tasks for each activity, (2) the adoption of a protocol and a set of templates to record and share information between activities, and (3) the implementation of a set of tests to be applied during the methodology phases.

5.1   Case Studies

In this section we describe our experience with two real SBI projects, which helped us in tailoring our methodology and demonstrating that an engineered approach positively impacts on the project success, meant in terms of both correctness and productivity. In particular we analyze two projects: a level-1 project in the subject area of Italian politics (PR-Pol) and a level-2 project in the subject area of a large consumer goods company (PR-CG). Both projects adopted an iterative approach and the tasks carried out are approximately the same, but while in PR-Pol our methodology was enforced, in PR-CG the team was mainly guided by its previous experience. As shown later, this leads to some inefficiencies in PR-CG.

The PR-CG working group was led by a system integrator with significant skills in SBI, featuring one project manager, one chief of consulting services, and six developers. The team was completed by an external scientific supervisor and by the innovation chief of the customer company. Though PR-CG was a level-2 project, we had a chance to monitor the activities of both the customer and the system integrator. The PR-Pol working group was quite smaller: it only included one project manager, one scientific supervisor, two developers, and the customer (the mayor of a large Italian city in this case). Overall, though the two projects are not fully comparable in terms of size and working group composition, they cover most of the critical issues related to SBI projects so they provide a good support for discussing the features of our methodology.

According to the classification proposed by [21], our case studies can be described as *explanatory/exploratory* (they aim at confirming the effectiveness of our methodology in real contexts, but also at finding new insights and at better tuning the approach), *positivist* (they use effort and correctness measurements), *quantitative* and *qualitative* (they quantitatively assess the validity of the approach, but they also collect qualitative judgments by the team), and *flexible* (due to the inherent dynamics of an SBI project, the requirements continuously change during the case studies). A more complete description can be given by answering the basic questions proposed by [22]:

- *Objective—What to achieve?*: the case studies aim at proving that the adoption of our methodology has a positive impact on the productivity and correctness of SBI projects.
- *The case—What is studied?*: we study two real projects with different characteristics and in different areas; both projects were carried out by skilled teams but with different compositions and size.
- *Theory—Frame of reference*: the theoretical framework we adopted is the one defined by the activities and tasks our methodology builds upon.
- *Research questions—What to know?*: we study how the two projects differ in terms of required effort and delivered utility.
- *Methods—How to collect data?*: for PR-CG, the effort for the different activities and tasks was derived a posteriori from an analysis of the time-sheets recorded by the system integrator. For PR-Pol it has been measured at project time. As to correctness, it has been estimated by asking some domain experts to manually tag a set of clips and comparing the resulting tags with those automatically obtained by semantic enrichment.
- *Selection strategy—Where to seek data?*: we selected two projects of different levels to achieve a wide coverage of the aspects involved in SBI design. PR-Pol was a level-1 project on a very wide and dynamic domain, led by a small team; PR-CG was a level-2 project on a more narrow domain, led by a system integrator.

In Table 3 we show the time spent on each task distinguishing the first iterations from the maintenance ones; missing items in the maintenance column denote activities made on demand, i.e., only at some iterations. Some comments on the values reported are necessary:

- Even if macro-analysis poses no particular problems, it usually requires a large amount of time because it is carried out during non-technical meetings that involve several different corporate departments.
- Maintaining the domain ontology requires more time in PR-Pol than in PR-CG. The reason is that the Italian politics subject area is quite wider than the consumer goods one, which implies a larger amount of dynamic contents to be analyzed in order to verify which new topics are to be added to the ontology.

Table 3. Time spent on tasks, expressed in man-days for first iterations and in man-days per week in maintenance iterations (n.a. stands for not available because the task has been outsourced)

Activity/Task	PR-CG		PR-Pol	
	1st Iter.	Maint. Iter.	1st Iter.	Maint. Iter.
Macro-Analysis	10	—	9	—
Ontology Design	4	0.6	7	1.5
Topics Definition	2	0.5	2	1
Inter-Topic Relation Def.	2	0.1	5	0.5
Source Selection	3	1	5	1
Semantic Enrichment Design	7	0.75	5	1
Crawling Design	10	1	29	1.5
Template Design	n.a.	n.a.	15	—
Query Design & Cont. Rel. Analysis	10	1	14	1.5
ETL & OLAP Design	15	—	24	—
ETL Design & Implem.	5	—	10	—
KPI Design	5	—	7	—
Dashboard design	5	—	7	—
Execution & Test	3	—	5	—
Total	52	3.35	84	5
In charge to the customer	15	0.85	84	5

- The time saving in semantic enrichment design for PR-Pol is mainly due to the adoption of a structured set of tests that has led the team to easily obtain the desired level of performances. This time saving is not apparent in maintenance iterations due to the higher complexity of the politics subject area.
- In query design and content relevance analysis, the amount of time needed to test how the developed queries work largely depends on the project level. In a level-2 project, the customer usually delegates crawling to an external service provider, who normally is capable of estimating the volume of clips retrieved by each specified query. Conversely, in a level-1 project, crawling has to be managed in every aspect, so that the effectiveness of a query can only be assessed after a whole clip acquisition session, that usually lasts 24 h; as a result, the execution of this activity can be significantly longer.
- The customer's effort is clearly reduced in a level-2 project. In particular, if no external provider is used for crawling, template design may end up for being very time consuming, which results in the largest time overhead.

As to semantic enrichment design, we focus on sentiment analysis, one of the more complex and important phases of the SBI process, that consists in determining the sentiment associated to a specific clip. Though the correctness of this analysis is obviously related to the capabilities of the semantic enrichment

Table 4. Correctness of sentiment analysis

	PR-CG			PR-Pol		
	Non-tuned	Tuned	Improvement	Non-tuned	Tuned	Improvement
Total	54.0 %	57.4 %	3.4 %	51.8 %	60.3 %	8.5 %
Social	52.5 %	55.9 %	3.4 %	46.1 %	58.1 %	12.0 %
Qualified	55.0 %	58.3 %	3.3 %	54.6 %	61.4 %	6.8 %
Hard	34.3 %	37.2 %	2.9 %	35.0 %	47.0 %	12.0 %
Standard	67.3 %	71.1 %	3.8 %	61.4 %	68.1 %	6.7 %
Negative	46.6 %	46.6 %	0.0 %	50.5 %	59.7 %	9.2 %
Neutral	45.6 %	49.1 %	3.5 %	62.0 %	71.3 %	9.3 %
Positive	69.5 %	76.3 %	6.8 %	47.8 %	52.4 %	4.6 %

engine[3], a fine tuning can lead to dramatic improvements. Both our projects share the same engine: *SyN Semantic Center*, a well-known commercial suite that enables a linguistic and semantic analysis of any piece of textual information based on its morphology, syntax, and semantics using logical-functional rules. So we investigated how the correctness of sentiment analysis was affected by the adoption of our methodology by asking five domain experts to manually tag a large set of clips (about 1,500) with their sentiment and then submitting them to the tuned/non-tuned engine. Tuning had a similar duration in the two projects (about two months) and led to a similar number of changes in the engine (about 330). Table 4 shows the results: clips are classified according to three criteria (media type, difficulty of a human expert in defining the sentiment, sentiment); the correct sentiment is assumed to be the one chosen by the majority of the domain experts. The semantic engine initially performed worse for Pr-Pol than for Pr-CG because the politics subject area uses a wider terminology and is probably more complex than the consumer goods one. However, the improvements obtained for Pr-Pol are clearly larger than those for Pr-CG. An in-depth analysis of the approach adopted by the Pr-CG team evidenced a lack of attention to the side effects of word polarization, that often introduced as many errors as those that were solved (see the extreme case of negatively polarized clips where no improvement is achieved.). Conversely, a more structured approach (see Semantic Enrichment design phase) and a continuous and iterative check of the side effects made the PR-Pol team's effort more effective.

Our case studies confirmed that ontology design and crawling design are the two most strictly-coupled activities and that their synchronization is a key factor to increase the overall performance. On the one hand, within crawling

---

[3] Evaluating sentiment analysis results is a difficult task since they may change a lot depending on the clip domain, the type of sources considered, etc. [23]. Nonetheless a reference value for sentiment analysis accuracy on simple domains is around 60 % − 70 %. Please consider, that a group of human experts typically find an agreement on the sentiment in the 80 % of cases.

design, the query design and content relevance analysis tasks are based on the topics determined by ontology design; on the other, the coverage achieved for the domain ontology mostly depends on how effectively crawling is able to exclude off-topic clips. In PR-Pol, at each iteration of ontology design, coverage analysis of the available clips is always made twice: once before adding new topics and once afterwords. The clips that remain uncovered are then handed on, together with the updated ontology, to crawling design and signaled as off-topic clips (i.e., crawling queries must be updated to discard these clips). This simple but effective protocol is applied every two days; in about 8 solar weeks the topics in the ontology increased from 139 to 225, and its coverage from 93 % to 98 %.

5.2   Case Studies Outcomes and Discussion

Responsiveness in an SBI project is not a choice but rather a necessity, since the frequency of changes requires a tight involvement of domain experts to detect these changes and rapid iterations to keep the process well-tuned. Such a frantic setting imposes a radical change in the project management approach with reference to traditional BI projects and a huge effort to both decision makers and developers (about one full-time person in both our projects). To reduce such effort, customers often try to outsource the activities yielding the worst trade-off between effort and added value for the SBI process. Besides the different technical skills required, this is the main motivation for conducting a level-2 project rather than a level-1 one.

During a project review session we analyzed, together with some members of the PR-CG team, the main problems they perceived, that turned out to be a lack of synchronization between the activities, that reduced their effectiveness, and an insufficient control on the effects of changes. With our methodology we tried to solve such problems through:

- A clear organization of goals and tasks for each activity.
- A protocol and a set of templates (not discussed in this paper for brevity) to record and share information between activities.
- A set of tests to be applied. The definition of each test includes the testing method and the indicators that measure the test results, for instance in terms of correctness of a process phase, as well as how these results have improved over the previous iteration.

6   Conclusions

In this paper we have discussed some of the key issues related to SBI. It is apparent that such original features requires specific techniques and methodologies to properly carry out a project in this area. In particular, the main feature that require to be addressed is the domain *dynamicity*. Both the meta-star modeling technique presented in Sect. 3 and the methodology presented in Sect. 5 address this issue.

Currently most of the implementations belong to the level 3 architecture and are carried out by third-party consultants such as web agency and digital marketing experts. This emphasizes that SBI systems are not considered an integral part of the information system yet. Keeping social data out of the information system does not allow the achievement of one of the SBI' main goal: the integration between Corporate (internal) Information and Social (external) ones. We finally remark that SBI is at the crossroad between different disciplines, this makes researches more challenging but it potentially opens to more interesting results.

References

1. Castellanos, M., Dayal, U., Hsu, M., Ghosh, R., Dekhil, M., Lu, Y., Zhang, L., Schreiman, M.: LCI: a social channel analysis platform for live customer intelligence. In: Proceedings of SIGMOD, pp. 1049–1058 (2011)
2. Rehman, N., Mansmann, S., Weiler, A., Scholl, M.: Building a data warehouse for twitter stream exploration. In: Proceedings of ASONAM, pp. 1341–1348 (2012)
3. Zhang, D., Zhai, C., Han, J.: Topic cube: topic modeling for OLAP on multidimensional text databases. In: Proceedings of SDM, pp. 1123–1134 (2009)
4. García-Moya, L., Kudama, S., Aramburu, M.J., Llavori, R.B.: Storing and analysing voice of the market data in the corporate data warehouse. Inf. Syst. Front. **15**(3), 331–349 (2013)
5. Gallinucci, E., Golfarelli, M., Rizzi, S.: Meta-stars: multidimensional modeling for social business intelligence. In: Proceedings of DOLAP, pp. 11–18 (2013)
6. Liu, B., Zhang, L.: A survey of opinion mining and sentiment analysis. In: Aggarwal, C.C., Zhai, C. (eds.) Mining Text Data, pp. 415–463. Springer, New York (2012)
7. Bravo-Marquez, F., Mendoza, M., Poblete, B.: Meta-level sentiment models for big social data analysis. Knowl. Based Syst. **69**, 86–99 (2014)
8. Pang, B., Lee, L.: Opinion mining and sentiment analysis. Found. Trends Inf. Retr. **2**(1–2), 1–135 (2008)
9. Lee, J., Grossman, D.A., Frieder, O., McCabe, M.C.: Integrating structured data and text: a multi-dimensional approach. In: Proceedings of ITCC, pp. 264–271 (2000)
10. Ravat, F., Teste, O., Tournier, R., Zurfluh, G.: Top_keyword: an aggregation function for textual document OLAP. In: Song, I.-Y., Eder, J., Nguyen, T.M. (eds.) DaWaK 2008. LNCS, vol. 5182, pp. 55–64. Springer, Heidelberg (2008)
11. García-Moya, L., Kudama, S., Aramburu, M., Berlanga, R.: Storing and analysing voice of the market data in the corporate data warehouse. Inf. Syst. Front. **15**(3), 331–349 (2013)
12. Dayal, U., Gupta, C., Castellanos, M., Wang, S., Garcia-Solaco, M.: Of Cubes, DAGs and hierarchical correlations: a novel conceptual model for analyzing social media data. In: Cheung, D., Ram, S., Atzeni, P. (eds.) ER 2012 Main Conference 2012. LNCS, vol. 7532, pp. 30–49. Springer, Heidelberg (2012)
13. Pang, B., Lee, L., Vaithyanathan, S.: Thumbs up? Sentiment classification using machine learning techniques. In: Proceedings of EMNLP, vol. 10, pp. 79–86 (2002)
14. Taboada, M., Brooke, J., Tofiloski, M., Voll, K.D., Stede, M.: Lexicon-based methods for sentiment analysis. Comput. Linguist. **37**(2), 267–307 (2011)
15. Pedersen, T.B., Jensen, C.S., Dyreson, C.E.: A foundation for capturing and querying complex multidimensional data. Inf. Syst. **26**(5), 383–423 (2001)

16. Gallinucci, E., Golfarelli, M., Rizzi, S.: Advanced topic modeling for social business intelligence. Inf. Syst. **53**, 87–106 (2015)
17. Kimball, R., Ross, M.: The Data Warehouse Toolkit: The Definitive Guide to Dimensional Modeling, 3rd edn. Wiley, New York (2013)
18. Golfarelli, M., Rizzi, S.: Data Warehouse Design: Modern Principles and Methodologies. McGraw-Hill, New York (2009)
19. Golfarelli, M., Lechtenbörger, J., Rizzi, S., Vossen, G.: Schema versioning in data warehouses: enabling cross-version querying via schema augmentation. Data Knowl. Eng. **59**(2), 435–459 (2006)
20. Francia, M., Golfarelli, M., Rizzi, S.: A methodology for social BI. In: Proceedings IDEAS, pp. 207–216 (2014)
21. Runeson, P., Höst, M.: Guidelines for conducting and reporting case study research in software engineering. Empirical Softw. Eng. **14**(2), 131–164 (2009)
22. Robson, C.: Real World Research. Blackwell, Oxford (2002)
23. Mullen, T., Collier, N.: Sentiment analysis using support vector machines with diverse information sources. In: EMNLP, vol. 4, pp. 412–418 (2004)

Context-Aware Business Intelligence

Rafael Berlanga$^{(\boxtimes)}$ and Victoria Nebot

Universitat Jaume I, Campus Riu Sec s/n, 12071 Castellón, Spain
{berlanga,romerom}@uji.es

Abstract. Modern business intelligence (BI) is currently shifting the focus from the corporate internal data to external fresh data, which can provide relevant contextual information for decision-making processes. Nowadays, most external data sources are available in the Web presented under different media such as blogs, news feeds, social networks, linked open data, data services, and so on. Selecting and transforming these data into actionable insights that can be integrated with corporate data warehouses are challenging issues that have concerned the BI community during the last decade. Big size, high dynamicity, high heterogeneity, text richness and low quality are some of the properties of these data that make their integration much harder than internal (mostly relational) data sources. In this lecture, we review the major opportunities, challenges, and enabling technologies to accomplish the integration of external and internal data. We also introduce some interesting use case to show how context-aware data can be integrated into corporate decision-making.

Keywords: Business Intelligence · Context-awareness · External data · Linked open data

1 Introduction

In the context of highly dynamic and global business scenarios, companies are engaged in the pursuit of new indicators that can provide them competitive advantages. Currently, the Business Intelligence (BI) community is paying special attention to exploiting the new massive data sources that are irrupting in the Web (e.g., social networks, sensors, data services, etc.) in order to define new breakthrough indicators and predictors for business. So far, this great effort is being carried out without taking into account the existing business models with which companies define their strategic goals and actions (e.g., [1]). On the other hand, BI analytics has been traditionally confined to corporate data, mainly gathered from internal IT processes, paying little attention to the business contextual information. Web-derived massive data sources open now new opportunities to automate the gathering of relevant contextual data and to integrate them with the BI analysis models.

Despite the great variety of commercial tools aimed at exploiting user generated data (e.g., trends in opinions about products or services), there is little work concerning the integration of relevant external data to corporate analysis so

© Springer International Publishing Switzerland 2016
E. Zimányi and A. Abelló (Eds.): eBISS 2015, LNBIP 253, pp. 87–110, 2016.
DOI: 10.1007/978-3-319-39243-1_4

that new context-sensitive indicators and predictors can be defined. For example, companies consider very valuable all published information implying some threat or opportunity to their business (e.g., some commerce legislation change or conflicts occurring at the providers' countries, changes in market trends, etc.) [2]. In this context, finding out correlations between detected external events and specific company's goals and indicators is a crucial task so that context takes part in the analysis. Nowadays, this is one of the most challenging issues for modern BI technology.

In this paper, we introduce the main concepts underlying a context-aware BI system. Then we review the main approaches in the literature aimed at context-aware BI, emphasizing their main strengths and limitations. We also discuss current trends and future research lines in this field. Finally, we introduce a use case to show the main aspects of a data infrastructure especially designed for context-aware BI.

2   Context-Aware Business Intelligence

BI is the process of collecting business data and turning it into information that is meaningful and actionable towards a strategic goal. Hence, BI technology is aimed at gathering, transforming and summarizing available business data from available sources to generate analytic information suitable for decision-making tasks. So far BI systems have been confined to corporate internal data, paying little attention to context information. The context of a BI system comprises all relevant external events and facts that could affect somehow the strategic goals of a company [1,3]. Therefore, a context-aware BI system should be able to properly monitor and produce actionable data from the business context, as well as to find relevant correlations to the company strategic goals. For example, insurance companies are mainly concerned with events that potentially affect their key indicators such as weather events, fraudulent incidences, etc. Moreover, companies are very interested in finding good indicator predictors when these events occur.

Although the BI definition implicitly implies both internal and external data, current BI technology mainly focuses on internal corporate data (i.e., Data Warehouse and OLAP technologies). Unfortunately, internal and external data are radically different in their nature, which makes them quite difficult to integrate under the same technological platform.

In this section, we discuss the layered structure of the business context, and the current technology concerning the extraction and publication of context-derived data.

2.1   Context Layers

In order to categorize the external data that is relevant for a BI system, we propose a layered organization based on the proximity and potential impact to the company business. Figure 1 shows these context layers, which are described in turn.

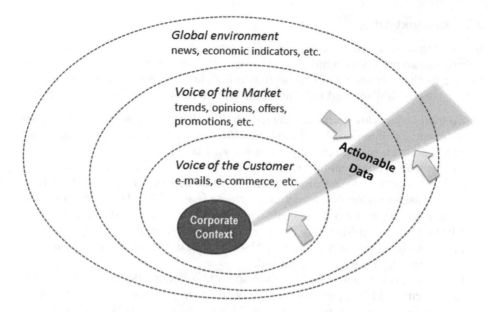

Fig. 1. Context Layers for BI

The most inner layer covers all events and facts that are internally produced in the company (e.g., sales, contracts, offers, etc.) This is the context where traditional data warehousing (DW) and OLAP take place. The second layer is the Voice of the Customer (VoC), which comprises all the events and facts directly produced by the clients or customers. Facts in this context usually refer to satisfaction indicators as well as opinions about the company's products or services. The third layer is the Voice of the Market (VoM), which mainly comprises relevant events and facts generated by competitors and potential customers. Facts and events in this layer are more heterogeneous than in the VoC context. Here, companies can identify market trends, global opinions in the same sector, impact of product promotions, etc. Finally, the global environment layer covers the rest of events that can indirectly affect the business indicators (e.g., situations in a business model [1]). Global trends in economy as well as any kind of news related to the company business can potentially affect its figures. Clearly, this context layer is the most noisy and difficult to manage within BI systems due to its high heterogeneity.

Currently, context data can be easily accessed through different web media. Blogs, social networks, news feeds and the web of data (i.e., data services) are the most outstanding examples of contextual data sources. In its whole, managing all these data sources is indeed a true Big Data problem, which involves its well-known four V's: velocity, volume, variety and veracity.

2.2   Context Objects

In order to manage context data in a BI system, first we have to identify their nature and structure. A context object is an external event or fact that we want to link to the corporate indicators. Regarding the data sources from which we gather context objects and their aims, we can identify the following properties:

- Context objects are multidimensional, that is, they can be naturally described by means of dimensions and measures. Moreover, there are two compulsory dimensions for all context objects: space and time.
- Context objects are mainly extracted from text-rich web data, such as documents, reports, blogs, opinion posts, and so on. Despite the fact that they are published under semi-structured formats (e.g., XML, RDF, etc.), relevant elements defining facts and events are usually expressed in natural language within text-free fields. This issue is partially alleviated by the presence of metadata (e.g., publication date, geolocalization data, etc.), from which basic contextual data are easily extracted.
- Context objects are usually expressed at different detail levels (multi-granularity). That is, events can imply different temporal or spatial extensions. For example, some events can affect a whole country while others a specific city.
- Context objects are usually incomplete and imprecise. That is, some relevant dimensions could be missing from extracted context data because either they are not reported or they are implicitly expressed.
- Context objects are usually typified (e.g., topics). For example, events could be classified into "natural disasters", "crisis events", etc., depending on the situations that analysts want to capture in their business models. In this way, classification is a way to abstract heterogeneous events into useful categories from the company point of view.

Some examples of context objects extracted from textual data sources are shown in Table 1.

Table 1. Example of context objects extracted from different sources

"Last floods in Spain caused losses of 1 million euros to the food company X". @News **(location:"Spain", time:"1/10/1998", company:"X", damage:"1 million euros")**
"Y because is a great car!" @Twitter **(location:"CS", time:"1/10/2015", product:"Y", opinion:"+1")**
"I didn't like movie Y :-(" @Twitter **(location:"BCN", time:"1/10/2015", product:"Y", opinion:"-1")**

Notice that some properties of context objects are well supported by traditional DW data models, like multidimensionality and multi-granularity. However, data quality is usually much poorer than in internal corporate data due to incompleteness and vagueness of event descriptions. Consequently, context objects are difficult to be allocated under the same data structures as corporate data. This is why most BI-aware systems aim instead at correlating context objects with DW objects, so that BI analysis can be properly contextualized. Moreover, the distinction between dimensions and measures for context objects (if possible) should be defined when performing the DW correlation for a particular analysis task.

In Table 2 we emphasize the main differences of internal and external data sources, which makes their correlation a difficult task.

Table 2. Main differences between internal and (web-based) external sources

Internal sources	External sources
Slow changes	Highly dynamic
Relational data	Un-/Semi- structured
High quality data	Low quality data
Complete information	Incomplete information
Historical data	Fresh data (real time)

2.3 Extracting Context Objects

Once relevant web-based data sources are located, context objects need to be extracted from web documents in order to correlate them with internal data.

For text-rich data sources the extraction must rely on natural language processing (NLP) techniques. More specifically, information extraction (IE) is the sub-field of NLP concerned with the extraction of structured records from texts written in natural language [4]. IE techniques rely on either manually specified extraction patterns or statistically inferred extraction models. In both cases, human intervention is essential and therefore their scalability is quite limited. In order to achieve web-scale performance, open information extraction (OIE) has been proposed [5]. Basically, OIE methods are self-supervised and rely on the availability of massive data. Although their former aim was to extract simple triple patterns of the form (subject, verb, object), these methods can be applied to other patterns. The main limitation of OIE is that these methods do not provide semantic information about the found triples, which is necessary to correlate context and corporate data.

A promising field for identifying and extracting context objects is the automatic semantic annotation (SA) [6]. In this case, we assume the existence of large knowledge resources (KRs) that serve as reference for annotating context objects with the entities they contain. Nowadays there exist several large KRs such as Wikipedia, BabelNet, BioPortal, etc. with a great coverage of many application domains. The main idea of SA-based methods consists in detecting occurrences of KR entities in the target data source so that plausible records can be generated with the detected annotations. The main advantages of these methods are that they do not necessarily need to perform any expensive NLP processing (e.g., POS-tagging, dependency analysis, etc.) and, thanks to the KRs, semantic annotations can be normalized through unique entity identifiers. It must be pointed out that automatic SA can also be applied to semi-structured and structured records gathered from web data services and micro-data.

2.4   Two-Level Context Correlations

Let us assume that we want to check if an external event E and a corporate fact F are correlated (i.e., they can be merged or aligned for integration purposes). For this purpose we need to perform a two-level test:

- The **first-level correlation** consists in checking whether the context surrounding E is relevant to the context surrounding F.
- The **second-level correlation** consists in checking the compatibility between the shared dimension values of both objects E and F.

As an example, let us consider the following corporate fact of a rent-a-car company

$$F = (location : ``CS", time : ``1/10/2015", product : ``Y", sales : ``1567")$$

Only the second external fact of Table 1 is deemed relevant to the company topics. Additionally, this external fact can be correlated to the corporate fact through the values of the shared dimensions (location, time and product).

Regarding the first correlation level several techniques coming from the Information Retrieval and Text Mining areas have been applied. Context correlation can be seen as the measurement of the vocabulary overlap between the external and internal contexts. Internal contexts can be set up from the data warehouse schemas and metadata [7]. External contexts are usually built from the document from which facts and events are extracted. IR provides numerous methods to capture the relevance of external contexts for a given internal context, namely: space vector models, language models, relevance models, topic models, etc. All of them basically consist in defining a weighting scheme for all vocabulary terms (indexing), and then comparing the resulting document representations based on this scheme.

Text Mining (TM) has been also applied to infer the implicit topics that can be identified in the contexts at hand [8]. Basically, a topic is represented as a statistical distribution of terms. Given a set of relevant topics to a given context, documents are represented as a mixture of topics according to the terms they contain. Comparing contexts is then reduced to comparing topic distributions. Several statistical inference techniques have been proposed to infer topics from text collections (e.g., Latent Dirichlet Allocation, Hierarchical Dirichlet Processes, and many variants of them), which can be directly applied to correlate internal and external contexts.

Checking the relevance of contexts can be also seen as a classification problem, namely: given an external context we classify it as relevant or not with respect to company goals. Unlike IR and TM methods, classifiers require human annotated examples of relevant and non-relevant external contexts (i.e., they are supervised methods). Although classifiers are often more precise than IR and TM methods, they do not work well in open scenarios like the Web.

Regarding the second correlation level, traditional information integration techniques are applied. Basically, we want to check whether two data records can be consistently merged. This task corresponds to the well-known record linkage problem (also known as entity recognition, data fusion and schema matching) [9], which concerns with proper similarity metrics that allow determining if two distinct data values are referring to the same entity. More recently, this problem has been extended to the integration of ontological resources, giving rise to the ontology alignment/merging field [10]. Figure 2 summarizes the techniques employed in the literature to perform the two-level correlation of context and corporate data.

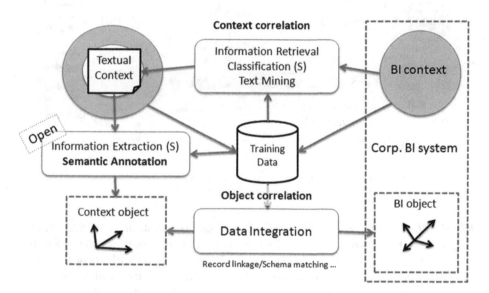

Fig. 2. Techniques and processes for context correlation. Supervised approaches are marked with (S).

3   Context-Aware BI Systems

Despite the fact that some work in the DW literature has attempted to define multi-dimensional models for non-conventional data [11], we cannot consider them as context-aware BI systems. A context-aware BI system is intended to identify and correlate relevant external data to the corporate analyses. Therefore, they can be seen as an extension of the traditional DW workflow where external information can be integrated on-demand at any phase. Figure 3 sketches an on-demand BI architecture, where a new process for extracting, transforming and querying (ETQ) external data is included [12]. Basically, the idea is that users are able to fetch information requests in order to enrich their analysis with relevant external data. These external data can be used in a variety of tasks: to find explanation of some indicator's change, to find correlations between external and internal indicators (e.g., opinions vs. sales), to predict indicators' evolution, etc. In this section we are going to review some context-aware BI systems that somehow follow the architecture shown in Fig. 3.

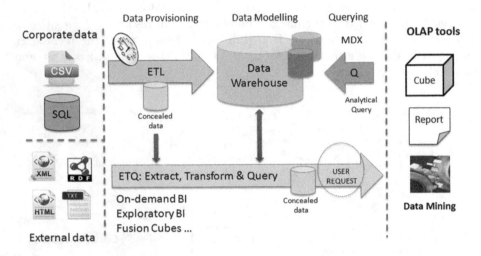

Fig. 3. BI architecture to account for context external data

EROCS/LIPTUS [13,14]. The goal of this pioneer system is to find out the corporate data record (e.g., a sale fact) that best correlates to a given document (e.g., email). In this work, customer e-mails are correlated with corporate facts. Context objects consist of a data record containing the entities identified in the document. As customer emails are assumed to talk about the company products, no context correlation is needed (first level correlation). Context and corporate objects are compared by using an adaptation of the TF*IDF metric [15] applied to the matched values. As a result, EROCS outputs a link table that associates each document to one corporate record. This table allows OLAP operations to be safely performed over the resulting integrated cube (i.e., IBM Alpha Cubes).

R-Cubes [16]. In this work, corporate data are correlated with news published in the Web. More specifically, traditional OLAP analysis is extended with context information by querying an XML document warehouse populated with news. Once the user fetches a query, relevant documents are retrieved and processed to extract the intended context objects. Similar to EROCS, in R-Cubes context objects are extracted by identifying corporate entities in the news documents. Unlike EROCS, this approach allows many to many relationships between news and corporate facts. The main contribution of this work is that both context and object correlations are calculated through well-grounded IR relevance models [17], which give a sound statistical foundation to the correlation measurement. Moreover, the properties of these statistical models allow OLAP operators to be safely applied to the relevance and context dimensions of the integrated cubes (i.e., R-Cubes).

SIE-OBI [18]. The goal of SIE-OBI is to find multidimensional correlations between a stream of contract documents (internal data) and a stream of news (external data). The first level correlation is performed by a classifier, which decides whether a news document is relevant or not for a given contract. Context and corporate objects are extracted from documents by applying a trained information extraction system. Finally, the correlation between these objects is performed with the distance in a space of Hierarchical Neighborhood Trees (HTN). In a HTN space, object values expressed at different detail levels (multigranularity) can be compared.

Opinion Cubes [7]. In this work, authors propose to build a data warehouse of opinion facts extracted from streams of product reviews (external data). The corporate and sentiment data warehouses share some key dimensions (i.e., time, place and product), which allow corporate cubes to be contextualized with opinion cubes by just joining them. The first-level correlation is achieved through tailored controlled vocabularies from both corporate metadata and Wikipedia relevant categories. Opinion facts are then extracted from product reviews by semantically annotating texts with the previous controlled vocabularies. The second-level correlation is performed by matching the dimensions shared by the corporate and sentiment data warehouses.

SLOD-BI [19]. The goal of SLOD-BI is to externalize the generation and publication of context objects to the Web of Data. Following the standards and formats of the Web of Data, extracted context objects are represented as linked open data [20]. As a result, a data infrastructure is provided to the analysts so that they can contextualize their corporate analyses. Context objects stem from different sources and they can be generated with different extraction methods. For example, published opinion facts could be directly extracted from texts by applying semantic analysis, or be parsed from micro-data embedded in the opinion posts. Whatever the extraction method is applied, all facts follow the same patterns and are normalized according to the existing controlled vocabularies provided by the infrastructure.

Comparison of Systems. As previously mentioned, all the previous approaches follow somehow the ETQ principle. Except for the SLOD-BI approach, the other systems only deal with one external data source. Regarding the covered context layers depicted in Fig. 1, none of the approaches cover all the layers (see Table 3).

Table 3. Coverage of context layers

Approach	VoC	VoM	Envir.
EROCS	yes	no	no
R-Cubes	no	no	yes
SIE-OBI	yes	no	yes
Opinion Cubes	no	yes	no
SLOD-BI	yes	yes	no

According to the techniques for extracting context objects (see Table 4), non-supervised are preferred to supervised ones mainly due to the open nature of external data.

Table 4. Two-level correlation methods

Approach	Context correlation	Object correlation
EROCS	-	Semantic annotation
R-Cubes	IR relevance models	Entity recognition
SIE-OBI	Classifier	Information Extraction
Opinion Cubes	Tailored vocabularies	Semantic Annotation
SLOD-BI	Linked Open Data (LOD)	Semantic Annotation

We can also compare these systems according to the classification criteria proposed in [12], namely: materialization, transformations, freshness, structuredness and extensibility. Most of the ETQ approaches are placed near conventional DW/OLAP technology, that is: full materialization, complex transformations, periodic refresh (ETLing) and limited extensibility (static data sources). Regarding the reviewed approaches, only the management of semi- and un-structured data makes the difference with respect to conventional methods. Clearly, the characteristics of these systems are far from the desired properties of modern BI, namely: virtual materialization, lightweight transformations, data stream processing (i.e., fresh data), no structural constraints on data, and dynamic extensibility with new data sources. The use of open data infrastructures like in SLOD-BI allows these systems to improve the transformation and extensibility issues. However, full freshness and extensibility are not yet achieved by current BI technology.

4 Context-Aware BI Use Case

The use case selected to demonstrate the feasibility of context-aware BI is the car rental domain. At the core of each rent-a-car company lies the idea of providing their customers cost effective and quality services. This vision must be reflected on each of their business activities, which range from accepting new reservations for new and existing customers to selecting the best promotional offer plan for each customer or managing the fleet, among others.

To ensure business success, companies often have a series of strategic goals, such as optimum utilization of resources, customer satisfaction or controlling costs, which are materialized by more specific and measurable objectives. The objectives are set up as the result of a decision making process, which usually involves complex analytic queries over corporate data. The most established approach is to use a DW to periodically store information subject to analysis. In the case of a rent-a-car company, the DW schema to analyze rental agreements includes typical analysis dimensions such as the rented vehicles, locations, customer features, etc. In order to make decisions, analysts often request the generation of reports and charts involving analytic queries, e.g., number of rental agreements per location and time or preferred rented vehicles by location.

Apart from traditional analytic queries involving corporate data, there is a need to get more insight of the business processes in real time to be able to react more efficiently (e.g., active data warehouses). In particular, customer satisfaction has become the greatest asset to success and there is a growing need of knowing customers' opinions about the companies' products and services. The consolidation of the Web 2.0 and the proliferation of opinion blogs and social networks has made available massive amounts of sentiment data subject to analysis.

In order to dynamically integrate corporate data with relevant social data to analyze the answer of customers to the company's strategic goals and to predict the demands of the market we propose the context-aware BI architecture for the car rental domain shown in Fig. 4.

In this architecture, corporate data are placed in the center and are loosely-coupled with external data sources by means of a semantic middleware. As external sources for our use case, we consider several car blogs, news feeds related to the car industry and twitter. The key of the architecture is the semantic middleware. All semi- and un-structured text-rich social data is converted to the semantic middleware format, so that other services can consume data from there and integrate it with corporate data.

For the semantic middleware, we propose an open and semantic infrastructure based on LOD (i.e., SLOD-BI), whose representation language is RDF. To give shared and well-defined meaning to the entities in the semantic middleware, we link them to external KRs such Dbpedia, Babelnet, etc. These KRs allow us to unambiguously identify entities in the texts and also to share and re-use the data. The process of identifying and linking natural language expressions to entities in such KRs is performed using automatic SA. For example, given the following opinion: *"The Mazda 5 has a useless backseat for anyone with legs"*, the SA tool

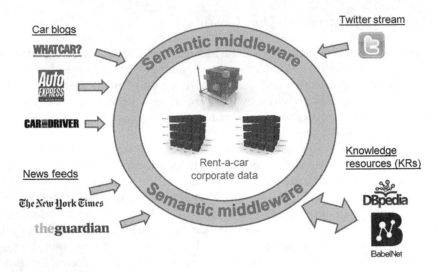

Fig. 4. Proposed architecture for the car rental use case

is able to identify the text chunk `Mazda 5` and link it to the Babelnet resource identifier http://babelnet.org/synset?word=bn:02438084n, which is the concept referring to the Mazda 5 car.

4.1 SLOD-BI Patterns

Social data is expressed in the semantic middleware in terms of multidimensional patterns (i.e., context objects), which can later be correlated with the corporate data model (i.e., corporate objects) for helping decision making. For example, the reputation of a product, the most outstanding features of some brand, or the opined aspects of an item can be represented as multidimensional data, and efficiently computed through OLAP tools [7]. Figure 5 summarizes the main multidimensional e-commerce patterns to analyze and correlate corporate and social data. The analysis patterns at the corporate data side of the figure (left part) correspond to the traditional multidimensional model of a typical DW. However, the patterns of the social data (right part) need some explanation as they constitute our proposal of analysis model for sentiment data extracted from relevant social data.

In the figure, facts (labelled with F) represent spatio-temporal observations of some measure (e.g., units sold, units offered, number of positive reviews, and so on), whereas dimensions (labelled with D) represent the contexts of such observations. In some cases, facts can have a dual nature, behaving as either facts or dimensions according to the analysis at hand. For example, a post can be either a fact or a dimension of an opinion fact. Dimensions can further provide different detail levels (labelled with L). For example, the dimension Item is provided with the level Sentiment Topic. In the figure, we distinguish two kinds of corporate

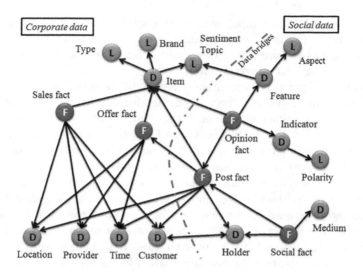

Fig. 5. SLOD-BI patterns

facts that can be correlated with social data, namely: corporate facts concerning business transactions (e.g., sales, contracts, etc.), and market facts concerning promotions and offers of the company.

The main facts concerning social BI are opinion facts, post facts, and social facts. Opinion facts are observations about sentiments expressed by opinion holders concerning concrete facets about an item, along with their sentiment indicators. For example, the sentence *"I dont like the dashboard of this car"* expresses an opinion fact where the facet is *"dashboard"*, and the sentiment indicator is *"dont like"* (negative polarity). Post facts are observations of published information about some target item, which can include a series of opinion facts. Examples of post facts can be reviews, tweets, and comments published in a social network. Finally, social facts are observations about the opinion holders that interchange sentiments about some topic. The latter facts are usually extracted from social networks by analyzing the structure emerged when the opinion holders discuss about some topic. Notice that topic-based communities can be very dynamic as they rise and fall according to time-dependent topics (e.g., news, events, and so on).

As for the measures associated to these facts, we can make use of typical measures in the literature for sentiment and social analysis, such as the polarity for opinion facts, usually expressed as positive, negative or neutral, the stars rating system or the number of likes for post facts, and measures such as the popularity or credibility for social facts.

It is important to notice that in Fig. 5 the corporate and social BI patterns are separated by data bridges, which are patterns that can be used to execute analysis operations that combine corporate and social data. For example, the analytic pattern between market facts and post facts can be applied to study the features of marketing campaigns from the point of view of its acceptance by consumers, and in the other way, to analyze consumer opinions in the context of each campaign. Different applications and different scenarios can make use of different data bridges to integrate data.

Data bridges support the communication channels between the internal and external data sources and it is very important for companies to enable all the means necessary to implement them. Of special interest are the data bridges that relate Sentiment Topics to Facets, and Customer to Holder dimensions. In the first case, the company can specify the most important topics in its items (products or services) that require some sentiment analysis. These topics are usually expressed as facets in the opinions of a post. In order to facilitate the implementation of this data bridge, companies and social media users could apply the same hashtags to mark up these topics. In the second case, it is important to note that when the holder of an opinion is a known customer, both entities must be identified as the same. With respect to these data bridges companies must ensure that the corporate data and metadata files include key information to enable the recognition of corporate entities in social data by means of sentiment analysis tools.

4.2   Examples of SLOD-BI Patterns

Figure 6 shows an excerpt of an opinion about a car that has been extracted from a textual review and linked to the SLOD-BI infrastructure. The textual review is converted to a `PostFact` and has several properties such as the item that is being reviewed, the reviewer, the body of the review, etc. Notice that all facts concerning the reviewers (e.g., popularity, number of posts, and so on) are expressed as `SocialFact`, which are not treated in this demo. An `Item` is any product or service subject to review. In our use case items are cars. Items have properties attached such as the text label, the domain the item belongs to, the brand of the product, etc. An `OpinionFact` expresses an association between a facet of an item and opinion indicators that appear at the post text. Thus, an opinion fact is always linked to the post object from which it was identified. Opinion facts have also attached a polarity, which is a numeric property that summarizes the overall sentiment of the opinion fact. A `Facet` of an item is any element subject to evaluation in the user's opinions, whereas a `SentimentIndicator` is

Fig. 6. Example of text opinion converted to the SLOD-BI infrastructure

a set of words that express some opinion about a subject together with their polarity (positive, negative, neutral). In the example, the opinion fact is relating the facet *interior* of a car with the sentiment indicator *attractive*, and has an overall polarity of *4*.

4.3 Structural View of SLOD-BI

Regarding the organization of the data in the semantic infrastructure devised, Fig. 7 proposes a structure for the intended social BI data. The involved datasets are divided into two layers. The inner layer regards the main vocabularies and datasets of the proposed infrastructure, whereas the outer layer comprises the external linked open vocabularies (LOV), and the KRs that are directly related to the infrastructure (e.g., Dbpedia, ProductDB, BabelNet, etc.). Every component consists of a series of RDF-triple datasets regarding some of the perspectives we consider relevant for BI over sentiment data. For example, in the Item component each dataset holds the products associated to a particular domain (e.g., cars, domestic devices, etc.) These datasets are elaborated and updated independently of each other, and can be allocated in different servers. All the datasets of a component share exactly the same schema (i.e., set of properties), which reflects the BI patterns defined in Fig. 5.

Links between components of the inner layer are considered hard links, in the sense that they must be semantically coherent, and they are frequently used when performing analysis tasks. On the other hand, links between infrastructure components and external datasets are considered soft links, as they just establish possible connections between entities of the infrastructure and external datasets. These external datasets are useful when performing exploratory analyses, that is, when new dimensions of analysis could be identified in external datasets. Links to external datasets like Dbpedia play a very relevant role in this infrastructure since they can facilitate the migration of existing review and opinion data. For example, reviews already containing micro-data referring to some product in Dbpedia will be automatically assigned to the product URI of the corresponding SLOD-BI product dataset.

4.4 Functional View of SLOD-BI

Figure 8 summarizes the functional view for the proposed data infrastructure. At the bottom layer, the external web data sources are selected and continuously monitored to extract, transform and link (ETLink) their contents according to the SLOD-BI infrastructure. As earlier stated, social BI facts (i.e., context objects) are regarded as spatio-temporal observations of user sentiments in social media. Therefore, both spatial and temporal attributes must be captured and explicitly reflected in the ETLink processes.

The SLOD-BI infrastructure is exploited by means of the data service layer, which is in charge of hosting all the services consuming sentiment data to produce the required data for the analytical tools. These services are implemented on top of a series of basic services provided by the infrastructure, namely: a SPARQL

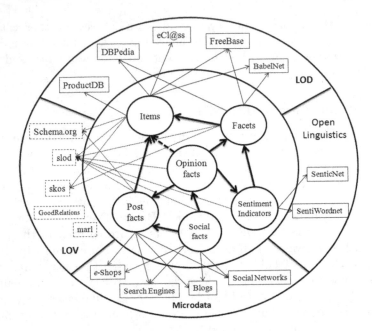

Fig. 7. Structural view of SLOD-BI

endpoint to directly perform queries over sentiment data, a Linking service to map corporate data to the infrastructure data (e.g., product names, locations, etc.), an RDF dumper to provide parts of the SLOD-BI to batch-processing services, an API for performing specific operations over the infrastructure (e.g., registering, implementing access restrictions over parts of the infrastructure, etc.) and visual tools for data exploration.

Notice that in the proposed functional view, sentiment data is integrated with corporate data using external analytical tools, by making use of some intermediate data service. In this case, corporate and sentiment data are aggregated separately and joined inside the analytical tools through a cross-join. This process is similar to Pentaho blending processes to integrate external and internal data[1]. The predictive models and exploration tools will allow the execution of complex processes over the sentiment data in the infrastructure.

5   SLOD-BI Demo

In this section, we show some specific examples of the SLOD-BI infrastructure. First, we show by means of several visual interfaces, some of the tools that enable the conversion of text data to the SLOD-BI infrastructure and also, some visual queries to the linked data in the infrastructure. Then, we show an example of integration of corporate data with SLOD-BI data using an external analysis tool.

---

[1] http://www.pentaho.com/big-data-blend-of-the-week.

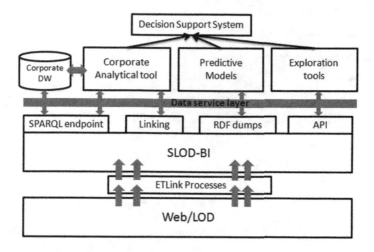

Fig. 8. Functional view of SLOD-BI

5.1   Tools and Visual Queries in SLOD-BI

For demonstration purposes, we have setup a website[2] where we show some of the services and tools used to convert social text data to the semantic middleware in the SLOD-BI infrastructure. Figures 9 and 10 show a screenshot of our tool for semantic annotation and the sentiment analysis tool, respectively. In the website, we also provide a SPARQL endpoint to query all the RDF data in the infrastructure.

In the same website, we can also access a visual interface of the SLOD-BI data, where opinion data about cars is queried, aggregated and displayed visually in the form of charts. For example, Fig. 11 shows in a bar chart the aggregated polarity of all the aspects belonging to the *comfortability* category for each of the cars. As social data is transformed to RDF data and arranged into multidimensional patterns (see Fig. 6), the SPARQL query to obtain this chart consists of a grouping of the data by *car* and by *sentiment topic* , a selection of the *comfortability* sentiment topic, and an aggregation of the *polarity* using the average function.

Figure 12 shows in a bar chart the aggregated polarity of all aspects by sentiment topic of the Ford Mondeo. The SPARQL query consists in a selection of the *Ford Mondeo* car, followed by a grouping by *sentiment topic* and an aggregation of the *polarity* using the average function.

Figure 13 shows a ranking of the highest and lowest scored aspects, along with their sentiment topic category, that are aggregated in the previous Fig. 12.

---

[2] http://krono.act.uji.es/EBISS/.

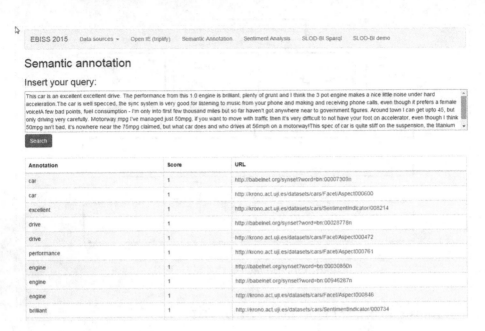

Fig. 9. Screenshot of the semantic annotation tool.

Fig. 10. Screenshot of the sentiment analysis tool.

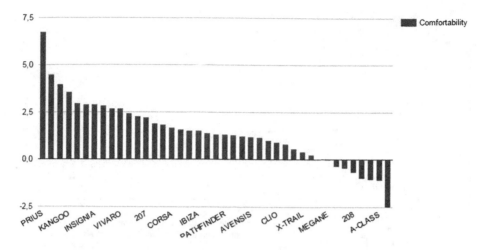

Fig. 11. Aggregated polarity of comfortability aspects of each car.

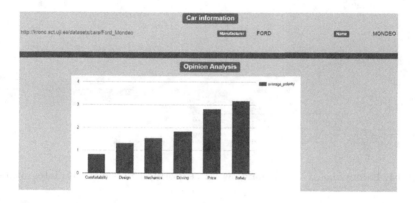

Fig. 12. Aggregated polarity of the aspects by sentiment topic of the Ford Mondeo

Top scored aspects				Less scored Aspects		
Aspect	Category	Score		Aspect	Category	Score
fan	Mechanics	9		visibility	Driving	-7.5
computer	Design	9		safety	Safety	-7
insurance	Safety	7		pedal	Mechanics	-6
money	Price	7		mirror	Mechanics	-4.333333333333333
mpg	Mechanics	5.7		reliability	Safety	-3.5
strong	Mechanics	5.125		window	Mechanics	-3.2
acceleration	Mechanics	5		clutch	Mechanics	-3
sound	Design	4.636363636363636		brake	Mechanics	-2.714285714285714

Fig. 13. Highest and lowest scored aspects of the Ford Mondeo car.

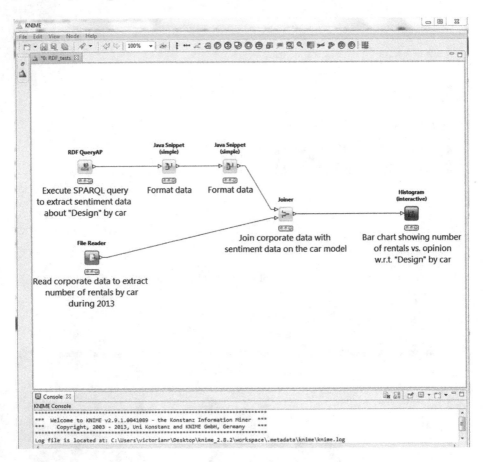

Fig. 14. Example of KNIME workflow for corporate and social data integration

5.2 Integration of Corporate and External Data

For the purpose of integrating corporate data with social data in the SLOD-BI infrastructure, we use the analytic tool KNIME[3]. KNIME is an open source data analytics, reporting and integration platform with a graphical user interface that allows the assembly of nodes for data pre-processing, modeling, analysis and visualization. We have implemented a node for performing SPARQL queries on SLOD-BI, and then we have used the workflow nodes of KNIME to integrate the social and corporate data. The resulting workflow is shown in Fig. 14. The bottom node queries corporate data to extract the number of rentals by car during 2013. The result is a table with two columns, the car and the number of rentals. The RDF QueryAP node executes a SPARQL query over the data service layer of the SLOD-BI infrastructure to extract sentiment data about the *Design* sentiment topic of cars. After some processing, the Joiner node merges

---

[3] http://www.knime.org.

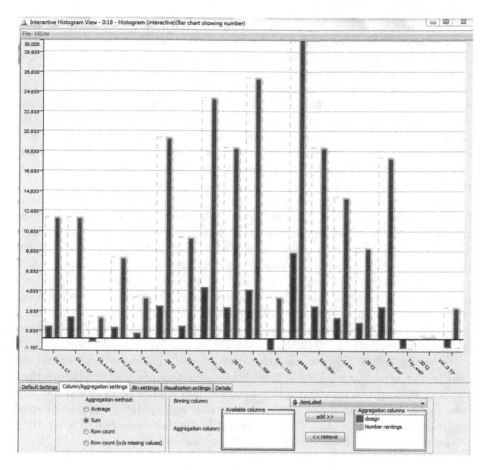

Fig. 15. KNIME integration results of corporate and social data (Color figure online)

the two tables on the car column and the resulting chart (Fig. 15) displays the number of car rentals (in blue) vs. the aggregated opinion on *Design* aspects (in red) by car. In general, we observe a positive correlation between the two variables, as the most rented cars (i.e., Renault Megane, Peugeot 208 and 508) are the ones with highest ratings on design aspects.

6 Challenges and Issues

As shown in previous sections, a context-aware BI system must be able to integrate relevant context data sources with the corporate analyses. We have shown a use case where several heterogeneous and dynamic data sources are plugged into a linked data infrastructure where context objects are published and consumed by the corporate analysis tools. Although this approach covers most of the desirable properties of a context-aware system (see Sect. 3), there are some challenging issues that should be addressed in the future work.

Firstly, context data is highly dynamic by nature and they are published in a streamed way. That is, context data are generated, published and disseminated across the different media as soon as they occur. However, existing approaches follow the batch processing approach of traditional data warehousing, which does not seem appropriate in this scenario. Therefore, the intended data infrastructure for capturing context objects should behave as linked data streams rather than static datasets. Streaming linked data is a recent topic addressed by the LOD community (e.g., [21,22]) and is mainly concerned with query languages and interchange formats. Additionally, the adoption by the industry of more flexible formats like JSON-LD can favor the massive generation of streamed linked data from the original data sources (e.g., social networks, news feeds, etc.)

Methods involved in the automatic extraction, annotation and classification of context objects (see Sect. 2.3) can be also affected by the streamed nature of context sources. As these methods rely on machine learning techniques, which require a training dataset to learn the models, they are subject to the concept drift problem [23], namely: the learned models can become obsolete as soon as the new arriving data cannot be explained by them. Clearly, unsupervised approaches can be better adapted to this situation as they do not need labeled training examples (and therefore continuous human intervention). Continuous learning is another challenging issue which is being addressed by the machine learning community (e.g., [24,25]).

Regarding to the BI concerns, there are several challenges that should be faced in the near future. Firstly, new business indicators and predictors that take into account both context and corporate data should be explored. On the other hand, the definition of new indexing and aggregation capabilities over graphs and linked data (e.g., [26,27]) should be further investigated since they are the main formats context objects are being published.

Acknowledgements. This work has been funded by the Spanish Economy and Competitiveness Ministry (MINECO) with project contract TIN2014-55335-R.

References

1. Horkoff, J., Barone, D., Jiang, L., Yu, E.S.K., Amyot, D., Borgida, A., Mylopoulos, J.: Strategic business modeling: representation and reasoning. Softw. Syst. Model. **13**(3), 1015–1041 (2014)
2. Thompson, J.L., Martin, F.: Strategic management: Awareness & change. Cengage Learning EMEA (2010)
3. Meredith, R., O'Donnell, P.: A functional model of social media and its application to business intelligence. In: Proceedings of the 2010 Conference on Bridging the Socio-technical Gap in Decision Support Systems: Challenges for the Next Decade, Amsterdam, The Netherlands, pp. 129–140. IOS Press (2010)
4. Sarawagi, S.: Information extraction. Found. Trends Databases **1**(3), 261–377 (2008)
5. Banko, M., Cafarella, M.J., Soderland, S., Broadhead, M., Etzioni, O.: Open information extraction for the web. IJCAI **7**, 2670–2676 (2007)

6. Uren, V., Cimiano, P., Iria, J., Handschuh, S., Vargas-Vera, M., Motta, E., Ciravegna, F.: Semantic annotation for knowledge management: requirements and a survey of the state of the art. Web Semant. Sci. Serv. Agents World Wide Web **4**(1), 14–28 (2006)
7. García-Moya, L., Kudama, S., Aramburu, M.J., Berlanga, R.: Storing and analysing voice of the market data in the corporate data warehouse. Inform. Syst. Front. **15**(3), 331–349 (2013)
8. Aggarwal, C.C., Zhai, C.: Mining Text Data. Springer Science & Business Media, New York (2012)
9. Koudas, N., Sarawagi, S., Srivastava, D.: Record linkage: Similarity measures and algorithms. In: Proceedings of the 2006 ACM SIGMOD International Conference on Management of Data, SIGMOD 2006, pp. 802–803. ACM, New York (2006)
10. Pavel, S., Euzenat, J.: Ontology matching: State of the art and future challenges. IEEE Trans. Knowl. Data Eng. **25**(1), 158–176 (2013)
11. Pérez, J.M., Berlanga, R., Aramburu, M.J., Pedersen, T.B.: Integrating data ware-houses with web data: a survey. IEEE Trans. Knowl. Data Eng. **20**(7), 940–955 (2008)
12. Abelló, A., Romero, O., Pedersen, T.B., Berlanga, R., Nebot, V., Cabo, M.J.A., Simitsis, A.: Using semantic web technologies for exploratory OLAP: a survey. IEEE Trans. Knowl. Data Eng. **27**(2), 571–588 (2015)
13. Bhide, M., Gupta, A., Gupta, R., Roy, P., Mohania, M.K., Ichhaporia, Z.: LIPTUS: associating structured and unstructured information in a banking environment. In: Proceedings of the ACM SIGMOD International Conference on Management of Data, Beijing, China, June 12–14, 2007, pp. 915–924 (2007)
14. Bhide, M., Chakravarthy, V., Gupta, A., Gupta, H., Mohania, M.K., Puniyani, K., Roy, P., Roy, S., Sengar, V.S.: Enhanced business intelligence using EROCS. In: Proceedings of the 24th International Conference on Data Engineering, ICDE 2008, 7–12, April 2008, Cancún, México, pp. 1616–1619 (2008)
15. Baeza-Yates, R., Ribeiro-Neto, B., et al.: Modern Information Retrieval, vol. 463. ACM Press, New York (1999)
16. Pérez-Martínez, J.M., Berlanga, R., Aramburu, M.J., Pedersen, T.B.: Contextual-izing data warehouses with documents. Decis. Support Syst. **45**(1), 77–94 (2008)
17. Croft, B., Lafferty, J.: Language Modeling for Information Retrieval, vol. 13. Springer Science & Business Media, New York (2013)
18. Castellanos, M., Gupta, C., Wang, S., Dayal, U., Durazo, M.: A platform for situ-ational awareness in operational BI. Decis. Support Syst. **52**(4), 869–883 (2012)
19. Berlanga, R., García-Moya, L., Nebot, V., Aramburu, M.J., Sanz, I., Llidó, D.M.: SLOD-BI: an open data infrastructure for enabling social business intelligence. IJDWM **11**(4), 1–28 (2015)
20. Bizer, C.: The emerging web of linked data. IEEE Intell. Syst. **24**(5), 87–92 (2009)
21. Fernández, J.D., Llaves, A., Corcho, O.: Efficient RDF interchange (ERI) format for RDF data streams. In: Mika, P., et al. (eds.) ISWC 2014, Part II. LNCS, vol. 8797, pp. 244–259. Springer, Heidelberg (2014)
22. Balduini, M., Della Valle, E., Dell'Aglio, D., Tsytsarau, M., Palpanas, T., Confalonieri, C.: Social listening of city scale events using the streaming linked data framework. In: Alani, H., et al. (eds.) ISWC 2013, Part II. LNCS, vol. 8219, pp. 1–16. Springer, Heidelberg (2013)
23. Wang, H., Fan, W., Yu, P.S., Han, J.: Mining concept-drifting data streams using ensemble classifiers. In: Proceedings of the Ninth ACM SIGKDD International Conference on Knowledge Discovery and Data Mining, KDD 2003, pp. 226–235. ACM, New York (2003)

24. Gaber, M.M., Zaslavsky, A., Krishnaswamy, S.: Mining data streams: a review. SIGMOD Rec. **34**(2), 18–26 (2005)
25. Calders, T., Dexters, N., Gillis, J.J.M., Goethals, B.: Mining frequent itemsets in a stream. Inf. Syst. **39**, 233–255 (2014)
26. Nebot, V., Berlanga, R.: Towards analytical MD stars from linked data. In: KDIR 2014 - Proceedings of the International Conference on Knowledge Discovery and Information Retrieval, Rome, Italy, 21–24 October, 2014, pp. 117–125 (2014)
27. Gallinucci, E., Golfarelli, M., Rizzi, S.: Advanced topic modeling for social business intelligence. Inf. Syst. **53**, 87–106 (2015)

Key Performance Indicators in Data Warehouses

Manfred A. Jeusfeld[1][(✉)] and Samsethy Thoun[2]

[1] University of Skövde, IIT, Box 408, Portalen, 54128 Skövde, Sweden
manfred.jeusfeld@his.se
[2] Pannasastra University, 184, Norodom blvd, Phnom Penh, Cambodia
samsethy@gmail.com

Abstract. Key performance indicators are widely used to manage any type of processes including manufacturing, logistics, and business processes. We present an approach to map informal specifications of key performance indicators to prototypical data warehouse designs that support the calculation of the KPIs via aggregate queries. We argue that the derivation of the key performance indicators shall start from a process definition that includes scheduling and resource information.

Keywords: Key performance indicator · Data warehouse · Business process

1 Introduction

Key performance indicators (KPI) evaluate the success of an organization or of a particular activity in which it engages (source: Wikipedia). They are used to continuously monitor those activities [1] in order to understand and control them. Deming [2] pioneered this field by statistically correlating independent process parameters to dependent performance indicators known as statistical process control (SPC). In SPC, the process parameters are kept in certain ranges such that the dependent variables such as KPIs or the product quality also remains in certain predictable ranges. These ideas were later also applied to software engineering [3], and to business process management [1]. Typical examples of KPIs are number of defects of a product, customer satisfaction with a service, the profit margin of a product, the percentage of deliveries before the promised delivery time, the machine utilization in a factory, and so forth. All these examples relate in some respect to an activity or to sets of activities. Moreover, they involve the interaction of multiple objects or subjects such as customers, employees, or machines.

In this paper, we investigate the relation of KPIs, data warehouses, and business process management. Specifically, we propose a guideline for deriving a prototypical data warehouse design from annotated KPI definitions, which

---

Part of the research was carried out while the second author was carrying out his master thesis project in the Erasmus IMMIT program at Tilburg University, The Netherlands.

E. Zimányi and A. Abelló (Eds.): eBISS 2015, LNBIP 253, pp. 111–129, 2016.
DOI: 10.1007/978-3-319-39243-1_5

themselves are derived from business process model fragments. This yields a top-down data warehouse design that supports the calculation of the KPIs via aggregate queries.

A data warehouse consists of multi-dimensional facts representing measurable observations about subjects in time and space. The subjects, time, and space are forming the dimensions, and the measures are representing the observations about the participating subjects. A data warehouse is essentially a large collection of measurements covering a certain part of the reality. In most cases, these measurements are about processes. If it were not, it would only provide a static account of objects in the reality. The key problem of this paper is *how to design the data warehouse from annotated KPI definitions such that the KPIs can be calculated by aggregate queries on the data warehouse.*

Another angle to KPIs is their summarizing nature. A KPI is not based on a single arbitrary observation but it aggregates a large number of observations about the same entities (or activities) to be statistically meaningful. The concept of an observation is the atomic building block of KPIs. Once the common properties of observations are set, one can start to collect such observations systematically and create the KPI on top of them. Different types of observations lead to different KPIs. So, given the definition of a KPI, what is the type of observations belonging to this KPI? KPIs can also be formed as expressions over other more simple KPIs. For example, the productivity of a process is the division of a KPI on the output of the process divided by a KPI on the resources used for producing the output. Such KPIs are called *derived KPIs*. Since their computation is simple once the part KPIs are computed, we shall focus on *simple KPIs* that are not defined in terms of other KPIs but that are defined in terms of sets of atomic observations of the same type.

2   Related Work

Key performance indicators quantify the performance of an organization or of its processes to achieve business objectives. In this chapter we view KPIs as used in conceptual modeling, in particular business process modeling, and in data warehousing.

2.1   Key Performance Indicators in Conceptual Modeling

Wetzstein et al. [1] investigate the definition of KPIs in the context of business process models, in particular from a service-oriented architecture perspective. Simple KPIs (called process performance metrics, PPMs) are the basis of more sophisticated, context-specific KPIs such as determining whether a customer has received the promised quality of service QoS (e.g. response time) can depend on the customer class and further parameters that we can view as dimensions of the KPI measurement. In their view a KPI is based on PPMs, a QoS definition, and a decision tree that determines whether a PPM measurement fulfills the QoS definition.

Strategic business modeling based on the Business Intelligence Model BIM [4] extends the goal modeling language i* by metrics linked to i* goals on the one side and tasks on the other side. The goals are monitored by the metrics and the tasks are the measures to achieve the goals. The metric interval is decomposed into performance regions (target, threshold/acceptable, worst value). The approach reminds of balance scorecards but extends it to the rich goal modeling language i*.

In software engineering, KPIs were introduced to manage the software development process [22], in particular in combination with the capability and maturity model CMMI [23]. Measurements such as the defect density in source code are used to control the software development process. Oivo and Basili's goal-question-metric (GQM) approach [24] provides an informal guideline on which metrics need to be monitored in order to assess that a certain goal (like improving the software quality) is reached. A quality goal is decomposed in a set of quality questions, which is itself decomposed into a set of quality metrics. The metrics are comparable to KPIs. Hence, the GQM approach allows to group KPIs by the goals of stakeholders. An agreement on goals allows to focus only on those KPIs that are needed to assess to which extent the goals have been reached. The GQM approach highlights that metrics (and thus KPIs) should not be mixed up with goals. Nevertheless, quality goals are often formulated in terms of KPIs such as the average cycle time of a certain process must be below a certain threshold.

Statistical process control (SPC) [25] was introduced by Deming [2] and others into the manufacturing domain as a tool to monitor the production and product quality. Specifically, it measures parameters and establishes statistical correlations between the parameters (called variables in statistics). The correlations between variables are translated into a set of equations for predicting values for dependent variables from independent variables. The idea is to control the independent variables (such as the quality of input materials) at early stages of the production process in order to guarantee that the dependent variables (such as product quality parameters) are within a desired interval. The variables in SPC are comparable to KPIs.

2.2   Data Warehouse Design and KPIs

A central issue in data warehousing is to design appropriate multi-dimensional data models to support querying, exploring, reporting, and analysis as required by organizational decision making. DW design has received considerable research attention. However, there are different methodological approaches proposed by the literature. Some approaches are data-driven in the sense that they aim at deriving facts and dimensions from the structures of operational sources that are usually represented as Entity Relationship Diagrams (ERD) or Unified Modeling Language (UML) diagrams. The outcome of this approach is a set of candidate facts or even data schemas, among which only relevant ones are selected to include in DW systems. For instance, Golfarelli et al. [5] proposed the DW design approach based on E/R scheme. Golfarelli and Rizzi [6] also developed a data-driven method for DW design based on Dimensional Fact Model.

Song, Khare, and Dai [7] developed the SAMSTAR method that is a semi-automated approach to generating star schema from operational source ERD. Although, the authors mentioned that the SAMSTAR method was both data-driven and goal-driven, this method is primarily data-driven because it derives star schema based on the structures and semantic of operational sources. Zepeda, Celma, and Zatarain [8] proposed a conceptual design approach consisting of two stages. The first stage is to generate multidimensional data structures from UML-based enterprise schema. The second stage is to use user requirements to select relevant schema. Moreover, the algorithm for automatic conceptual schema development and evaluation based on Multidimensional Entity Relationship Model (M/ER) was invented by Phipps and Davis [9]. Similarly, Moody and Kortink [10] proposed a methodology for designing DW schema based on enterprise models.

On the other hand, a goal-driven approach gives more relevance to user requirements in designing DW. Prakash and Gosain [11] present a requirement-driven data warehouse development based on the goal-decision-information model. In addition, Giorgini, Rizzi, and Garzetti [12] propose a goal-oriented requirement analysis for DW design in which the organizational goals are made explicit and decomposed into sub-goals and then the relationships among sub-goals and actors are identified and analyzed. Their method starts with identification of corporate goals (i.e., user requirements) and actors involved. The actor can be either a responsible persons or resources that are needed to accomplish the goal.

We focus on the conceptual design phase to provide a blueprint for lower level logical design that is consistent with the KPI definitions from which we start. Tryfona, Busborg, and Christiansen [13] developed the starER model for conceptual design of Data Warehouses and argued that DW design should be exposed to higher level so that it becomes more understandable, and easier to identify conceptually what are ingredients are actually needed in the DW. In addition, it is advisable not to use computer metaphors such as 'table' or 'field'.

Jones and Song [14] developed Dimensional Design Pattern (DDP) that assists designers to effectively determine commonly used DW dimensions. In this sense, the DDP framework consist of six classes of dimension domain, from which DW designer can choose specific dimension and attributes during the mapping process.

Moreover, an important issue in designing DW schema is additivity of facts. A fact is additive relative to a dimension if it is summarizable along that dimension. The importance of summarizability is discussed by Shoshani [15]. Horner, Song, and Chen [16] present a taxonomy of summary constraints that can be used for this purpose.

The other issue in designing DW schema is the choice between the various types of multidimensional data models, among which star schema and snowflake schema are most common in data warehouses. However, the most data warehouses use star schema for two important reasons. First, it is the most efficient design because less joint operations are required due to denormalized tables. Second, the star schema is supported by most query optimizers for creating an access plan that use efficient star join operations [17].

A study of data warehouse in connection with KPIs can be found in the triple-driven data modeling methodology presented by Guo et al. [18]. This methodology consists of four major stages: (1) goal driven stage, (2) data driven stage, (3) user driven stage, and (4) combination stage. During the first stage, business goals and KPIs are identified according to business subject area. The second stage is to obtain a data schema that supports the KPIs from the operational data sources. The third stage is to interview users in order to identify important business questions. The fourth stage is to check if the business KPIs can be calculated and questions can be answered by the obtained data schema. As indicated by its second stage, this methodology is primarily data-driven because the operational sources impose total constraints on the computation of KPIs. Moreover, the first stage is where KPIs have to be identified and the attributes needed to support these KPIs have to be determined. However, this methodology does not specify how to determine those required attributes as part of the DW data models. In other words, the practical steps to analyze the KPI structural definition are not provided. In addition, the generation of star schema is based on the data-driven method that was developed by Moody and Kortink [10].

Vaisman and Zimányi [21] propose a classification of KPIs along several dimensions. First, KPIs are classified with respect to the time span of observations (past, present, future). Second, they distinguish KPIs on inputs needed for a business results from KPIs about the business result and performance. Further, there are operational vs. strategic KPIs and qualitative (obtained by surveys, etc.) vs. quantitative. Multidimensional expressions (MDX) relate a KPI value to a KPI goal (expressed as thresholds or intervals).

In the sequel, we develop an informal guideline on how to create a data warehouse schema out of patterns found in business process models. The multidimensional character of the KPIs is excerpted from the products serving as inputs and outputs of the processes, the resources used in the processes, and time and location information. We also shall review the role of plans and schedules (compare to targets in BIM) in formulating KPIs.

3   Data Warehouses for Structuring Observations

A data warehouse manages multi-dimensional facts, where each fact constitutes an observation about the domain of interest, e.g. an enterprise. The structure of an observation is a tuple

$$(d_1, d_2, \ldots, d_k, m)$$

where d_i are dimension entities represented by their identifier and m is a measurement value, typically a number. The measurement value attribute is functionally dependent on the combination of dimension entities. For example, assume that we have the dimensions car, location, and time and the measurement attribute 'speed' for representing car speed observations. Then, the observation facts would look like

```
('Marys car','Skövde',2013-09-28T10:31:19,385)
('Johns car','Barcelona',2013-03-12T21:07:47,145)
```

As functional expressions, these observations can be represented as equations

```
speed('Marys car','Skövde',2013-09-28T10:31:19)=385
speed('Johns car','Barcelona',2013-03-12T21:07:47)=145
```

We learn from this example that the dimensions of the observation determine the circumstances under which the speed observation was made. The car parameter is representing an entity participating in the observation. Location and time are dimension entities that frequently occur in observations. Other than the car, they are not entities/objects of the real world but we can reify them to be entities. This reification is common in data warehouses by creating dimension tables where temporal and special dimension values get surrogate identifiers. The goal of this paper is to derive the dimensions for a simple KPI from a high-level specification for this KPI.

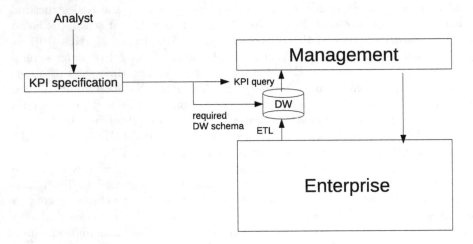

Fig. 1. Workflow of creating DW schemas from KPI definitions

The general steps for realizing the KPI are

1. Specify the KPI including its measurement context. The measurement context is defined by a combination of entities (customers, products, time, location, etc.) that were present when the observation was made.
2. Create the supporting data warehouse schema. We limit ourselves in this paper mostly on the fact table.
3. Code the queries computing the KPI on top of the created schema.

Natural language KPI definitions found in practice are usually rather ambiguous by nature. Take for example the average speed of cars as a KPI for the traffic process. What is the context of the underlying observations? It can be the time of the measurement, the location, and the car involved in the measurement. However, it could also include the car driver. Some of the relevant context may be difficult to determine such as the car driver. This can limit the utility of the KPI for decision making or for understanding the process underlying the observations.

3.1   The Process Nature of Observations

An observation is a statement made by agent (the observation) about an object in the reality, possibly involving other objects. Lenz and Shoshani [19] differentiate flow and stock observations. A *flow* observation is about recording a state change of the object recorded with respect to some time interval, a *stock* observation is a record about the object's state. As a third category, they list *value-per-unit* observations, such as the price of a product. Assume we would only record stock observations. If there are no changes, then the observations of an object would also not change. This is like listing the specific weights of elementary substances. If there are changes, then the states of objects vary over time and shall yield different observations. The reasons for changes are processes taking place in the reality. These processes can be natural like the radioactive decay or they are man-made, such as production processes. Consider the example of an oil refinery that stores oil in large tanks. Each tank has a fill level. There are two processes that can change the fill level: adding oil to the tank and removing oil. These processes are embedded in more complex processes taking place at the oil refinery. Flow observations about the oil tank record how much the state of an object has changed between two points of time. For example, how many liters of oil have been added and how many have been removed in the last month. If the state is known at the start of the time period, then the state at the end of the time period can be calculated by applying the additions and subtractions of the flow observations. The third observation type, value-per-unit break down stock or flow observations to small units, such as the oil price per liter. Assume that the oil refinery buys quantities of oil on the market at different prices and then stores the oil in the tank. Then each liter of oil stored in a tank virtually carries its unit price with it. The total value of the oil in the tank is then the sum of all oil liter unit prices of oil liters stored in the tank.

The lesson learned from this argumentation is that state changes require the presence of processes. If the processes are natural, then human influence on them is limited. For example, the water cycle on earth is driven by the sun and leads to varying levels of water in the river systems. Still, it makes perfect sense to record observations about the water cycle in order to predict the water levels of certain rivers at certain locations, e.g., to prepare for flooding. An organization with man-made processes has an interest in managing the processes to achieve its goals, e.g., to increase the profit or to raise customer satisfaction. The management includes changing the parameters of process steps (e.g., their scheduling), adapting the resources (e.g. the machines used in production steps), changing the inputs of process steps (e.g., replacing a part by another part), or changing the process itself (e.g., reordering the process steps or removing unnecessary activities).

A single observation occurs in a context, which is characterized by the participating entities. Time and space are regarded here as entities as well. The presence of time and space indicate that such observations are practically always related to an underlying process. The process is the reason why the entities are combined and lead to observations. As an example, consider the usage process of a customer c1 for a product p1 at a time t1. The observation for the combination of these three entities could be a defect of product p1. This is an atomic observation. The measurement attribute 'defect' is either 0 or 1.

Now, the customer c1 belongs to the set of customers, e.g., the set of customers in Brazil. The product p1 belongs to the set of all products of a given type, say ACME Phone-One. Then, we can state for example

```
Defects('Brazil','ACME Phone-One',2014) = 371
```

The example highlights that it is crucial to identify the context of an observation as a combination of participating entities. Combined with another simple KPI on the number of products of a given type sold in a country in a given year, one can define a derived KPI on the defect density:

```
DefectDensity('Brazil','ACME Phone-One',2014) =
    Defects('Brazil','ACME Phone-One',2014) /
    Sales('Brazil','ACME Phone-One',2014)
```

A *derived* KPI is simply a KPI that is defined in terms of other KPIs. A *simple* KPI is calculated from a set of atomic observations. Note that the arguments of the two KPIs 'Defects' and 'Sales' are the same, i.e., the context of the two underlying observation types is the same.

Figure 2 visualizes the step from multi-dimensional atomic observations (upper part) to multi-dimensional aggregated observations (lower part). The aggregated observations are sets of atomic observations about the usage activities of cutomers with products where the dimension entities of the atomic observations are member of the dimension values of the aggregated observation. The lower shows rolled-up dimension entities (all for customer, 2014 for time, and product group S). These dimension entities match a set of observations, which can be aggregated e.g. by counting the number of observations. Any set of atomic observations can define a multitude of KPIs by combining different dimension values. For example, the KPI

```
Defects('Brazil','ACME Phone-One',2014-01)
```

aggregates all defect observation in Brazil for the product group 'ACME Phone-One' in January 2014. We call all such KPIs simple KPIs even though equalities such as

```
Defects('All','ACME Phone-One',2014) =
    Defects('Brazil','ACME Phone-One',2014-01) +
    Defects('Brazil','ACME Phone-One',2014-02) +
    ...
    Defects('Brazil','ACME Phone-One',2014-12)
```

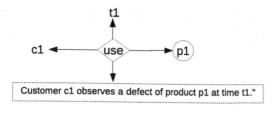

Customer c1 observes a defect of product p1 at time t1."

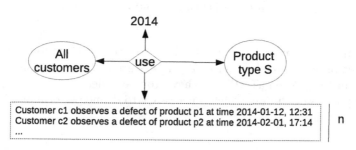

Customer c1 observes a defect of product p1 at time 2014-01-12, 12:31
Customer c2 observes a defect of product p2 at time 2014-02-01, 17:14
...

Fig. 2. Context of atomic and aggregated observations

hold true. The equality holds true due to the definition of the KPIs on the same set of atomic observation and the roll-up relations of the dimension entities.

We conclude that observations about processes are the basis to define KPIs and that the context of observations can be represented as a combination of entities such as products, resources, time, and location. These entities are the same entities that form the dimensions in a data warehouse. This view is not the only view on KPIs but it is the one used subsequently to create guidelines on how to derive data warehouse schemas and queries from KPI definitions.

4   From KPI Definitions to Data Warehouse Schemas

As motivated before, any KPI is based on observations about underlying processes. We focus on simple KPIs here, i.e. KPIs that are based on a single type of observation denoted as

$$O(e_1, e_2, \ldots, m)$$

where e_i are the entities participating in the observation and m is the value of the observation, usually a number. Hence, an observation is a synonym to a fact in a data warehouse where all dimension values are taken from the lowest rollup level. We also use the functional representation

$$O(e_1, e_2, \ldots) = m$$

when appropriate. Since the majority of KPIs are process-oriented, we use process models to relate them to elements of process models. Specifically, we

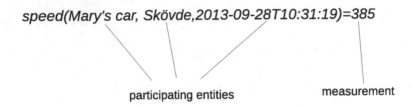

speed(Mary's car, Skövde,2013-09-28T10:31:19)=385

participating entities   measurement

Fig. 3. Participating entities and measurements

use the Petri net notation [20] extended by resource and input/output elements to represent process model patterns. Petri nets are the formal basis for process modeling languages such as BPMN. They provide a clear token passage semantics of the process execution, which is necessary to define performance indicators such as cycle time.

4.1 Motivating Example: Derive a DW Schema for the KPI "Average Speed of Cars"

Is it a simple or derived KPI? This is a simple KPI with an atomic underlying observation type.

What is the structure of the observation type? We identify the participating entities car (given by its identification), the location of the speed measurement, and the time when the measurement was taken. The measure is a number with unit km/h. Hence the type of the observation is

$$speed(CAR, LOC, TIM, SPEEDM)$$

What is the schema of the fact table of a data warehouse supporting the KPI? The participating entities become dimensions, e.g.,

```
CREATE TABLE SPEEDS (
    CARID INT,
    LOCID INT,
    TIMID INT,
    SPEEDM FLOAT,
    PRIMARY KEY (CARID,LOCID,TIMID),
    FOREIGN KEY (CARID) REFERENCES CAR (CARID),
    FOREIGN KEY (LOCID) REFERENCES LOCATION (LOCID),
    FOREIGN KEY (TIMID) REFERENCES TIMETBL (TIMID));
```

We omit the definitions of the dimension tables since the roll-up hierarchies are not mentioned in the KPI definitions. The query for computing the KPI is then a straightforward aggregate query on the fact table.

4.2   Pattern 1: Derive a DW Schema for the KPI "Average Processing Time for a Task in a Process"

Figure 4 shows a Petri-net-style process fragment to analyze the KPI. Place p1 represents that some case is currently being processed by task 1. The places are waiting positions for the cases that flow through the process. The two transitions 'begin' and 'end' start or terminate the task, respectively. A case is a data object representing an external or internal event to which an organization has to react. It carries an identifier (the case id) and possibly further attributes that describe the case. The attributes are used to decide how to route a case thru a process [20]. The *inner place* p_i is uniquely defined for each task in a process model.

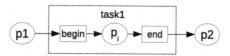

Fig. 4. Process fragment for understanding processing time

Is it a simple or derived KPI? This is a derived KPI based on the arrival and departure times of cases at the inner place p_i of a task.

What is the structure of the observation types? There are two observation types:

```
arrivaltime(CASE,PLACE,ARRTIME)
departuretime(CASE,PLACE,DEPTIME)
```

Here the time is not a participating entity but a measurement. There are two dimensions involved in the observation: the case dimension and the place dimension. The place dimension can be rolled up to the task to which it is connected and then to the process to which the task belongs.

What is the schema of the fact tables of a data warehouse supporting the KPI?

```
CREATE TABLE ARRIVALTIME (
    CASEID INT,
    PLACEID INT,
    ARRTIME DOUBLE,
    PRIMARY KEY (CASEID,PLACEID));
CREATE TABLE DEPARTURETIME (
    CASEID INT,
    PLACEID INT,
    DEPTIME DOUBLE,
    PRIMARY KEY (CASEID,PLACEID));
```

The query to compute the simple KPI `arrivaltime(o,p)` is then

```
SELECT ARRTIME FROM ARRIVALTIME WHERE
    CASEID = o AND
    PLACEID = p;
```

The two fact tables can also be merged into a single one with two measurement attributes. Foreign key references and the definitions of the dimension tables are omitted. The query to compute the KPI aggregates the average of the difference of the departure time of the inner place p1 of a given task. We leave the query coding to the reader. The Petri net view on the process allows to determine what events need to be recorded by a process execution system. For pattern 1, the system has to record the time when a case is picked up by a task (arrival time at p_i) and when the task finishes a case (departure time at p_i).

Fig. 5. Cycle time of process

4.3   Pattern 2: Average Cycle Time of a Case in a Process

The cycle time is the accumulated time of a case in a process, from start to end. Assume that ps is the unique start place of the process and pe is its unique end place, then the cycle time of a case c is a derived KPI based on the arrival time:

$$cycletime(c) = arrivaltime(c, pe) - arrivaltime(c, ps)$$

We thus can reuse the definition of $arrivaltime$ of the previous example. We assume that the process has a unique start ps and a unique end pe. The SQL query to compute the average cycle time over all cases is left to the reader. It multiple processes are analyzed by the same data warehouse, then one can add a process dimension to the fact table for $cycletime$. Processes can be rolled-up to process groups at the discretion of the data warehouse designer.

The cycletime is calculated here from the simple KPI $arrivaltime$. If the complete process definition is known, then one can establish an equality of the $cycletime$ with the sum of all waiting times plus all processing times for a case flowing through the process.

4.4   Pattern 3: Average Waiting Time on a Place

This is another derived KPI that can be defined in terms of arrival and departure time:

$$waittime(c, p) = arrivaltime(c, p) - departuretime(c, p)$$

The *waittime* can be aggregated to the total waiting time of a case in a process. If a process *proc* has no cycles, then it is defined by the formula

$procwaittime(proc) =$
$\quad sum\{arrivaltime(c, p) - departuretime(c, p) \mid c\ in\ CASE,\ p.process = proc\}$

If the process has cycles, then cases can visit the same place multiple times. Then, our original definitions for arrival and departure time cannot be used anymore. To solve the problem, we add an additional participating entity 'visit' that contains the identifier of the visit of a case on a place:

```
CREATE TABLE ARRIVALTIME (
    VISITID INT AUTOINCREMENT,
    CASEID INT,
    PLACEID INT,
    ARRTIME DOUBLE,
    PRIMARY KEY (VISITID,CASEID,PLACEID));
```

The fact table for departure time is updated accordingly. Then, the process waiting time can be defined as

$\qquad procwaittime(proc) =$
$\qquad\quad sum\{arrivaltime(v, c, p) - departuretime(v, c, p) \mid c\ in\ CASE,$
$\qquad\quad p.process = proc,\ v\ in\ INT\}$

We leave the SQL query for calculating the KPI to the reader.

4.5 Pattern 4: Average Person Hours Spent on a Task for a Given Case

Person hours are an example of a resource-based metric. Resources are allocated to tasks. They are reserved during the execution of the task and typically released before the end of the task. We can distinguish consumable resources such as energy and non-consumable resources such as machines or employees. The latter can be converted to consumable resources by considering resource hours instead of the resource itself.

Figure 6 links a resource to a task in a SADT-like style as also used by Fenton and Pfleeger for software processes [22]. In an SADT (structured analysis and

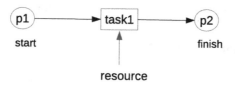

Fig. 6. Resources linked to tasks

design technique) diagram, a task has inputs, outputs, resources used for the task and control information (e.g. a time schedule). The resource consumption can be observed by identifying the current case, the task to be performed, the identifier of the resource. The measurement is the consumption of the resource, e.g. person hours. Hence the observation type is

$$personhours(CASE, TASK, RESOURCE, RHOURS)$$

The following fact table implements the observation type:

```
CREATE TABLE PERSONHOURS (
    CASEID INT,
    TASKID INT,
    RESOURCEID INT,
    RHOURS DOUBLE,
    PRIMARY KEY (CASEID,TASKID,RESOURCEID));
```

The query to compute the resource consumption per task and resource is then as follows:

```
SELECT TASKID, RESOURCEID, AVG(RHOURS)
FROM PERSONHOURS
GROUP BY (TASKID,RSOURCEID);
```

4.6   Pattern 5: Percentage of the Truck Shipment Time Where the Truck Cooling Device Is Active

This KPI is derived from the process time of the truck shipment and the aggregated cooling times of the cooling device resource. The first KPI is discussed in Example 2. Hence, we only need to handle the use of the cooling device.

$$cooling(ENGAGE, CASE, TASK, CTIME)$$

The cooling device can be engaged multiple times during a shipment. The observation has as participating entities the engagement id, the case, the task (ship) and as measurement the time of the engagement.

```
CREATE TABLE COOLING (
    ENGAGEID INT AUTOINCREMENT,
    CASEID INT,
    TASKID INT,
    CTIME DOUBLE,
    PRIMARY KEY (ENGAGEID, CASEID, TASKID));
```

The cooling time aggregated over all engagements for a given task and case is then:

```
SELECT TASKID, CASEID, SUM(CTIME)
FROM COOLING
GROUP BY (TASKID, CASEID);
```

In a similar way one can implement KPIs on power consumption of a machine resource used to perform a given task.

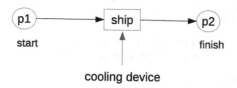

Fig. 7. Use of the cooling device

4.7   Pattern 6: Material Used to Create a Product

Physical processes create output products using input products. The input products are not resources but they become part of the output. To model KPIs for such processes, we need to explicitly represent inputs and outputs of tasks. Figure 8 shows the inputs and outputs of a task. Note that inputs and outputs of the task are different from the control flows from place 1 to the task, and from the task to place 2. A task can have multiple products as inputs and also produce multiple outputs. Each participating product can have a quantity (measured in physical units or as count). For example, to produce an engine for a car, one needs a certain quantity of aluminum poured in a form. The observation type for input products is characterized by the participating entities case, the task, and the input product. The measurement is the quantity of the product used for the task on the given case.

$$input(CASE, TASK, PRODUCT, QUANTITY)$$

The outputs can be characterized accordingly

$$output(CASE, TASK, PRODUCT, QUANTITY)$$

Let us assume that 23.23 kg of aluminum are used to create a certain engine 123. The observation facts would then be:

$$input(engine123, pour, aluminum, 23.23)$$
$$output(engine123, pour, engine, 1)$$

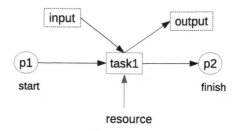

Fig. 8. Inputs and outputs of a task

The next engine could require slightly less aluminum:

$$input(engine124, pour, aluminum, 23.19)$$
$$output(engine124, pour, engine, 1)$$

```
CREATE TABLE INPUT (
    CASEID INT,
    TASKID INT,
    PRODUCTID INT,
    QUANTITY DOUBLE,
    PRIMARY KEY (CASEID,TASKID,PRODUCTID));
```

The average consumption of aluminum per engine is then a simple aggregate query over the input table:

```
SELECT TASKID, PRODUCTID, AVG(QUANTITY)
FROM INPUT
GROUP BY (TASKID,PRODUCTID);
```

4.8   Pattern 7: As-Is vs. To-Be Comparisons

The last pattern discussed in this chapter are deviations from the plan and KPIs that relate planned performance to the actual performance. A typical example is a budget for a project. This is a planned measure. The actual cost of the project may be less, equal, or more than the planned budget. Another example is the deadline for a certain task. The previous patterns already discussed the actual performance of a process, including resource consumption. Figure 9 adds planned performance to our extended process model. We can regard the planned performance as a simple observation type, which has no participating case.

As an example, consider the planned processing time of task 1. It can be represented in an observation fact

$$plannedproctime(TASK, PTIME)$$

This observation fact can be used like any other to form aggregate KPIs like the average planned processing time over all tasks. The more interesting use is to form derived KPIs with KPIs on the actual performance. Similar planned performance KPIs can be defined for resource consumption, and input and outputs.

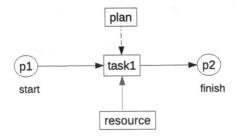

Fig. 9. Planned performance

5   Conclusions

This paper discussed how to map a KPI definition to a data warehouse schema and the query calculating the KPI. Rather than developing a method to automatically generate the schema and queries, we elaborated on patterns for process performance KPIs. The patterns included processing time, waiting time, resource consumption, material use, and the comparison of planned versus actual performance. An extended process model that includes places, tasks, resources, inputs/outputs, and plans was incorporated to derive the observation types underlying the KPIs. Simple KPIs have a single observation type associated to them. Derived KPIs are computed as expressions over simple KPIs.

The patterns can be used to support the top-down design of a data warehouse from a set of process-related KPIs that shall be computed by it. The starting points are the natural language KPI definition and a process model fragment that visualizes the context in which the observations belonging to the KPI are collected. The notion of Petri-net places allowed for a straight-forward definition of time-based KPIs by just using arrival and departure times of cases at and from places. The pattern on resource consumption allows dealing with a whole group of KPIs such as person hours spent on a task.

The input/output pattern allows to measure physical material flow. These patterns can also be combined with the other patterns, e.g. to measure how many person hours are needed to produce a certain number of products. Finally, planned performance is realized by a simplified observation type that has no case identifiers.

We argue that practically all KPIs are process-related because any change of a state requires some activity leading to the state change. Some KPIs are about 'stock' observations (cf. Lenz and Shoshani [19]), e.g., observing the number of cars on a certain street segment. The observation is related to the ongoing travel processes of the car drivers, which are not made explicit in an information system about the traffic status. The observation times are independent of the underlying travel processes: two consecutive observations could be about the very same state. If the process is not explicit, then it cannot be controlled so easily. 'Flow' observations are directly linked to a process task, since they make an explicit statement on a state change. In the traffic example, each time that a car enters or leaves the street segment, an observation would be recorded. This type of observation allows to control the traffic, e.g. by using traffic lights for the street segment that is set to red when too many cars are in the segment. In this paper, we thus focused on flow observations.

Future work is needed to understand how to define a KPI in a formal language such that a supporting data warehouse schema can be automatically generated from the KPI definition. Another open question is whether the discussed 7 patterns cover a considerable portion of KPIs actually used in practice. The KPI Library (http://kpilibrary.com) contains more than a thousand KPIs in high level natural language that can be used to answer this question. We did not discuss how dimension tables can be created and populated. Most rollup hierarchies are domain-specific with the exception of time.

Finally, it would be interesting to investigate rules for the correct definitions of derived KPIs in terms of constraints on the use of parameters for the participating entities of the observation facts. For example, it does not make (much) sense to compare the arrival times and departure times of places belonging to different processes.

KPIs can also be regarded as statistical variables, possibly depending on each other. The long-term collection of KPIs can be used to calculate their correlation and thus to form a theory on estimating dependent KPIs from independent ones. This paper was meant to encourage the systematic collection of many process KPIs such that theories for predicting them can be developed and validated using methods from SPC [25].

References

1. Wetzstein, B., Leitner, P., Rosenberg, F., Brandic, I., Dustdar, S., Leymann, F.: Monitoring and analyzing influential factors of business process performance. In: Proceedings of the 2009 IEEE International Enterprise Distributed Object Computing Conference, EDOC 2009, Auckland, New Zealand, pp. 141–150 (2009)
2. Deming, W.E.: On Probability as a basis for action. Am. Stat. **29**(4), 146–152 (1975)
3. CMMI Guidebook Acquirer Team: Understanding and Leveraging a Supplier's CMMI Efforts: A Guidebook for Acquirers, CMU/SEI-2007-TR-004. Software Engineering Institute, March 2007. http://resources.sei.cmu.edu/library/asset-view.cfm?assetID=8315
4. Horkoff, J., Barone, D., Jiang, L., Yu, E.S.K., Amyot, D., Borgida, A., Mylopoulos, J.: Strategic business modeling: representation and reasoning. Softw. Syst. Model. **13**(3), 1015–1041 (2014)
5. Golfarelli, M., Maio, D., Rizzi, S.: Conceptual design of data warehouses from E/R schema.In: Proceedings of the 31st Annual Hawaii International Conference on System Sciences, HICSS 1998, vol. 7, pp. 334–243 (1998)
6. Golfarelli, M., Rizzi, S.: Data Warehouse Design - Modern Principles and Methodologies. McGraw-Hill, Osborne (2009)
7. Song, I.-Y., Khare, R., Dai, B.: SAMSTAR: a semi-automated lexical method for generating star schemas from an entity-relationship diagram. In: Proceedings of the 10th ACM International Workshop on Data Warehousing and OLAP, DOLAP 2007, Lisbon, Portugal, pp. 9–16 (2007)
8. Zepeda, L., Celma, M., Zatarain, R.: A methodological framework for conceptual data warehouse design. In: Proceedings of the 43nd Annual Southeast Regional Conference, 2005, Kennesaw, Georgia, Alabama, USA, vol. 1, pp. 256–259 (2005)
9. Phipps, D., Davis, K.C.: Automating data warehouse conceptual schema design and evaluation. In: Proceedings of the 4th International Workshop on Design and Management of Data Warehouses, DMDW 2002, Toronto, Canada, CEUR-WS.org/Vol-58, pp. 23–32 (2002)
10. Moody, D., Kortink, M.A.R.: From enterprise models to dimensional models: a methodology for data warehouse and data mart design. In: Proceedings of the International Workshop on Design and Management of Data Warehouses, DMDW 2000, Stockholm, Sweden, CEUR-WS.org/Vol-28 (2000)

11. Prakash, N., Gosain, A.: Requirements Driven Data Warehouse Development. Short Paper Proceedings CAiSE 2003, Klagenfurt, Austria. CEUR-WS.org/Vol-74 (2003)
12. Giorgini, P., Rizzi, S., Garzetti, M.: Goal-oriented requirement analysis for data warehouse design. In: Proceedings of the 8th ACM International Workshop on Data Warehousing and OLAP, DOLAP 2005, Bremen, Germany, pp. 47–56 (2005)
13. Tryfona, N., Busborg, F., Christiansen, J.G.B.: StarER: a conceptual model for data warehouse design. In: Proceedings of the 2nd ACM International Workshop on Data Warehousing and OLAP, DOLAP 1999, Kansas City, Missouri, USA, pp. 3–8 (1999)
14. Jones, M.E., Song, I.-Y.: Dimensional modeling: identification, classification, and applying patterns. In: Proceedings of the 8th ACM International Workshop on Data Warehousing and OLAP, DOLAP 2005, Bremen, Germany, pp. 29–38 (2005)
15. Shoshani, A.: OLAP and statistical databases: Similarities and differences. In: Proceedings of the 16th ACM SIGACT-SIGMOD-SIGART Symposium on Principles of Database Systems, PODS 1997, Tucson, Arizona, USA, pp. 185–196 (1997)
16. Horner, J., Song, I.Y., Chen, P.P.: An analysis of additivity in OLAP systems. In: Proceedings of the 7th ACM International Workshop on Data Warehousing and OLAP, DOLAP 2004, Washington, DC, USA, pp. 83–91 (2004)
17. Martyn, T.: Reconsidering multi-dimensional schemas. ACM SIGMOD Rec. 33(1), 83–88 (2004)
18. Guo, Y., Tang, S., Tong, Y., Yang, D.: Triple-driven data modeling methodology in data warehousing: a case study. In: Proceedings of the 9th ACM International Workshop on Data Warehousing and OLAP, DOLAP 2006, Arlington, Virginia, USA, pp. 59–66 (2006)
19. Lenz, H.J., Shoshani, A.: Summarizability in OLAP and statistical databases. In: Proceedings of the 9th International Conference on Scientific and Statistical Database Management, SSDBM 1997, Olympia, Washington, USA, pp. 132–143 (1997)
20. van der Aalst, W.M.P., van Hee, K.M.: Workflow Management - Models, Methods, and Systems. MIT Press, Cambridge (2002)
21. Vaisman, A.A., Zimányi, E.: Data Warehouse Systems - Design and Implementation. Springer, New York (2014)
22. Fenton, N.E., Pfleeger, S.L.: Software Metrics - A Practical and Rigorous Approach. Thomson, Fenton (1996)
23. Software Engineering Institute: CMMI for Development - Version 1.3. Technical Report CMU/SEO-2010-TR-033, Carnegie-Mellon University, November 2010
24. Oivo, M., Basili, V.R.: Representing software engineering models - the TAME goal oriented approach. IEEE Trans. Softw. Eng. 18(10), 886–898 (1992)
25. Oakland, J.S.: Statistical Process Control, 6th edn. Routledge, New York (2008)

Author Index

Printed in the United States
By Bookmasters